Time, Cost, and Architecture

Time, Cost, and Architecture

George T. Heery, AIA

McGRAW-HILL BOOK COMPANY
New York St. Louis San Francisco Auckland Düsseldorf
Johannesburg Kuala Lumpur London Mexico Montreal
New Delhi Panama Paris São Paulo Singapore
Sydney Tokyo Toronto

Library of Congress Cataloging in Publication Data

Heery, George T
Time, cost, and architecture.

1. Building - Cost Control. 2. Construction Industry - Management. I. Title
TH437.H35 658.1'552 74-31456
ISBN 0-07-027815-6

1234567890 VHVH 78321098765

The editors for this book were Jeremy Robinson and Robert Braine, the designer was Naomi Auerbach, and the production supervisor was George Oechsner. It was set in Optima Medium by University Graphics, Inc. It was printed and bound by Von Hoffman Press, Inc.

Contents

Preface

It is possible for architects, engineers and construction managers to exert a highly acceptable degree of control not only over the cost of building construction but also over the time required for the design and construction process. Possibly as important but far less well recognized, it is feasible to predict the final cost of a building and its date of beneficial occupancy to a quite reasonable degree.

This book describes a definitive system for time and cost control that may be applied within any given program of requirements, quality level, or design goal. The system has been used effectively on many projects of varying size and type, for private and governmental owners, and in various areas of the United States and abroad, including some of the most difficult of construction markets.

The time/cost control system may be employed by the architect/engineer or by the separate professional construction manager. Both approaches are dealt with herein.

As the movement toward more and better management of construction programs has advanced and begun to formalize and as various systems have been proposed for better control of cost and time in building, these developments would appear to have taken place to some degree outside of the architectural and related engineering design professions.

Many would point out, and with some validity, that the evolution of construction management as a profession, or definable professional service, has taken place within and because of a management void left by the architectural and engineering professions.

Yet can something so intimately related to a building's design as the control of its cost and time of delivery be successfully separated from the design approach? It would seem clear that the two cannot be separated if that control is to be as effective as it can be.

Still another view is that when there is strong control of cost and time, the design must suffer. This kind of logic cannot stand close scrutiny, but nonetheless many architects and others seem to feel that it is a fact.

What, then, is the relationship between architectural design and construction management, and what should it be?

That relationship would seem to be, basically, that both are indigenous parts of the process that creates architecture, and that they both have to be carried out well and thoroughly intertwined to produce truly successful architecture.

It is hoped that this book will further contribute toward effective time/cost control through better design and construction management methods. It is also hoped that it will contribute toward a redirection of design thinking that will produce architecture that is not only visually successful but that also fulfills its social, economic, and environmental requirements.

Three types of projects are not dealt with in the book: single-family residences; the industrial process plant in which the cost is predominately process equipment, such as refineries and paper pulp mills; and heavy engineering projects such as highways, dams, etc. However, many of the principles contained herein are applicable, with variations, to these three project types, particularly the latter two.

I would like to thank and acknowledge the help of a number of my staff, associates, and fellow principals. Many hours of gathering illustrations, specimen specifications, photographs, and forms were contributed by Larry Dean, Bill Mitchell, Art Deas, Jim Fagerburg, Kitty Smith, and many others. Thanks to Will Ferguson for helping to pull it all together. And particularly, I acknowledge with great fondness and admiration, the help and encouragement over the years of my fellow principal, Vic Maloof, who helped develop the Time/Cost Control system. Many thanks to my father and senior partner, retired, who always let me do it my way, but who set awfully high standards. And, finally, thanks to Betty, who makes it all worthwhile, for her help and encouragement.

George T. Heery

CHAPTER 1

New Directions in Design with Construction Management: Fourteen Case Histories

For the architect, history is surely the most dependable point of reference. For any artist, a fresh, objective scrutiny of his medium provides a most dependable reorientation. For the construction manager, evaluation in terms of time and cost control within function and quality goals is the real test.

In the awesome, challenging, complex, and exciting last half of the twentieth century, the professional group that is made up of architects, engineers, construction managers, and related design professionals seeks something more than visual sensation. Surely evaluation of an end product must be based on a multifaceted criterion with history as the point of reference, time and cost control within established quality levels as the acid test, and above all, judgment on whether or not there has been a full understanding and creative utilization of the medium.

Time control within high standards of design and cost control may be the essence of expertise of professionals working with construction.

Yet all around us we see the product of the fashionable designer whose technology seems to have advanced only about as far as the Middle Ages. There seems to be little understanding or use of the available medium, and in too many cases there seems to have been a shunning of the new building's functional role, the owner's and society's timely need for the facility, and the owner's or taxpayer's pocketbook.

The foregoing is a damning criticism, true, and surely not deserved by many architects, but true indeed in too many instances, sometimes to a ludicrous point. So much of the work of those who are only fashion-conscious seems clearly irrelevant to society's needs, time control, cost control, and the medium of our time.

There are a number of distinguishable segments of the profession and the construction industry which have for some time been manifesting attitudes ranging from discontent with past methods to serious questions about design directions of the 1960s and 1970s. They include

- *Owners,* who have been more and more frustrated by the high costs, long design and construction times, and functional shortcomings (particularly as projects have become larger and more complicated during periods of inflation, labor problems, and growing bureaucracy). Current fashionable architecture has often been seen by these owners to come about through exclusively visual considerations and to be, therefore, often irrelevant to its proposed role.
- *Students,* who are often at odds with faculty and the profession alike. Many brighter students view with disdain, arising from a strong social consciousness, the "jewel box" building designed with visual and sculptural characteristics and the "Fountainhead" approach as its guiding force.

■ *Systems proponents,* a small but determined group in the profession, who have seen the great need to proceed toward industrialized building construction, or at least a systems approach to design for better control of time and cost.

■ *Construction management proponents,* who are attempting to focus on the problems of better control of time and cost and to bring better design and construction management, in harmony with one another, to bear on these problems.

■ *Minority labor groups* looking for better job opportunities and other industrious workers, both of whom, while looking for steadier employment, do not find an acceptable place today in the union-controlled construction industry with its low productivity.

■ *Certain practicing architects* and architectural/engineering firms that endeavor to be the complete practitioner by providing a competent construction program management service coupled with creative and functional design for their clients.

The project case histories in this chapter are reviewed not as optimum end products from either an architectural standpoint or from a construction management standpoint but rather because they represent some of the efforts of one professional group using a definitive design and construction management system for better time and cost control coupled with an integrated design.

They are not necessarily the most impressive projects of the abovementioned professional group (Heery & Heery of Atlanta and New York), but were selected as a cross section of type and size.

A most significant point about the illustrated projects is that they were all produced through combining architectural and engineering services with construction management services by an integrated professional group.

Just as the architects of Greece and Rome directly oversaw the building of their projects and fully understood the optimum use of the medium available to them, so must the late-twentieth-century architect be a complete architect by understanding the medium, utilizing the medium and the engineering technology available, employing modern management techniques, and bringing together other related design disciplines into a coordinated single entity of creative effort.

CASE HISTORY 1

FIFTEEN RECREATION CENTER/ SWIMMING POOL COMPLEXES FOR THE CITY OF NEW YORK PARKS DEPARTMENT

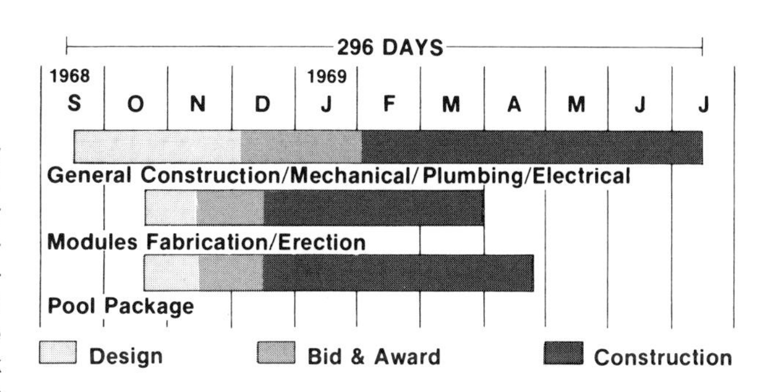

Total Construction Cost..$4,528,811

Under the administrations of Mayor John V. Lindsay and New York City Parks, Recreation and Cultural Affairs Administrator August Heckscher and later Administrator Richard Clurman, a series of fifteen recreation center/swimming pool complexes were undertaken. Sited in a number of different existing parks and other Parks Department land, the projects were located throughout the five boroughs of New York City. The design and construction schedule, as well as the cost figures shown here, were for the first six projects which were built during the spring and early summer of 1968, opening for use that summer. Previous similar projects of the Parks Department had taken two and a half to three years for design, administrative reviews, and construction.

In the predesign project analysis (see Chapter 5), by dealing simultaneously with several important design factors, the concept of a precast concrete module as the basic unit of an industrialized system was developed. In this analysis it was recognized that there would be a variety of sites with varying project sizes. Exposure to vandalism and rough use was expected. Availability of many of the sites was a major constraint in the scheduling of the projects. In turn, the fabrication and the erection of the building structures and enclosures were identified as potential constraints and had to be dealt with as critical activities. Consequently it was determined that, prior to the commencement of any project design and during the first forty-eight hours of the architectural and construction management work on the project, there should be a precast concrete module designed that would be substantially fabricated off site, that could easily be bid on by any number of existing precasters in the metropolitan New York area utilizing their existing facilities and familiar materials, and that could be moved easily through any of the tunnels and bridges leading to or within the City of New York. In this case, rather than utilizing performance specifications for the development of the system, the modules were fully designed in detail and prescriptive specifications were employed. In turn, the necessary quantities of the modules, along with contracts for their delivery and erection, were bid early (prior to the sites being available) along with the other major critical contract package, i.e., the swimming pool tanks and filter packages. (Aluminum tanks were utilized for the smaller pools; aluminum sidewalls and gutters were utilized in connection with concrete bottoms for the larger pools, and preassembled filtering units were specified, which presented a labor jurisdictional problem in only one of the smaller projects.)

Since these projects were public works subject to the laws of the State of New York, it was a requirement that there would be separate contracts for the mechanical work, plumbing (as differentiated from the other mechanical work), the electrical work, and the general construction. Because of these same laws, none of separate contracts could be transferred into the general contract. In turn, it was felt that there was no point in attempting to transfer the separate early contracts for the modules' fabrication and erection or the pool package work into the general contract.

All the construction scheduling was handled with the Critical Path Method as described in Chapter 9. Time-control construction-contract provisions basically similar to those discussed in Chapter 8 were incorporated into each contract. Architectural design, construction management, landscape architectural design, structural engineering, systems design, and civil engineering were all done by Heery & Heery. Mechanical and electrical engineering was done by S. A. Bogen & Associates of New York City under subcontract to Heery & Heery. Critical Path Method consultation services were provided by Management Science America under specifications and recommendations prepared by Heery & Heery.

east
harLem
pooL

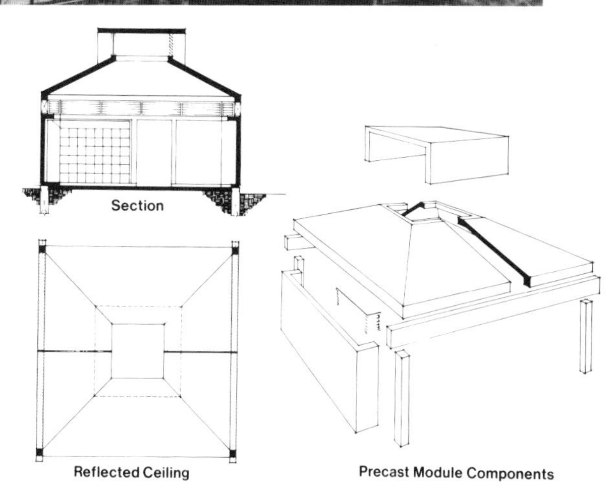
Section
Reflected Ceiling
Precast Module Components

CASE HISTORY 2

MANUFACTURING PLANT FOR THE KAWNEER COMPANY NEAR ATLANTA (JONESBORO), GEORGIA

This project was the first in which the time/cost control design and construction management system was utilized as a defined method. Because of the owner's desire for a lump-sum price before proceeding, this project may illustrate near the optimum in time control while still using competitive bidding and a single-responsibility general contract plan of construction management. The facility includes an office area of 3,936 square feet and a total plant size of 77,289 square feet. Plant operations include extruding, anodizing, and assembly of architectural aluminum products as well as warehousing and distribution. The cost of $4.13 per square foot in 1961 included all site work, finishes, mechanical and electrical work, and a 10-ton materials-handling bridge crane. The owner's representative, who had full authority to make decisions, "lived" in the architects' offices during the design phases and gave "on board" reviews and approvals.

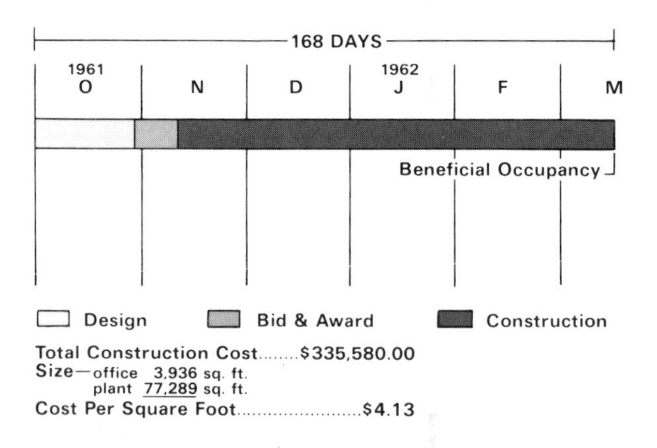

Total Construction Cost........$335,580.00
Size—office 3,936 sq. ft.
plant 77,289 sq. ft.
Cost Per Square Foot........................$4.13

The design program, prepared by the owner in advance of design, consisted primarily of an industrial engineering plant layout. This and a topographical survey were available to the architects at the outset.

In this time schedule, as in the case of all other projects' schedules, the term "design" refers to all phases of preliminaries and contract document preparation. "Beneficial Occupancy" refers to the full facility being ready for occupancy and full operation unless otherwise noted. Heery & Heery were architects and engineers and provided construction management under a single contract.

CASE HISTORY 3

ATLANTA STADIUM

Architectural and engineering services as well as construction management, for the Atlanta Stadium were provided to the owner, the Atlanta and Fulton County Recreation Authority, under a single contract by the joint venture of Heery & Heery and Finch, Alexander, Barnes, Rothschild & Paschal, both of Atlanta.

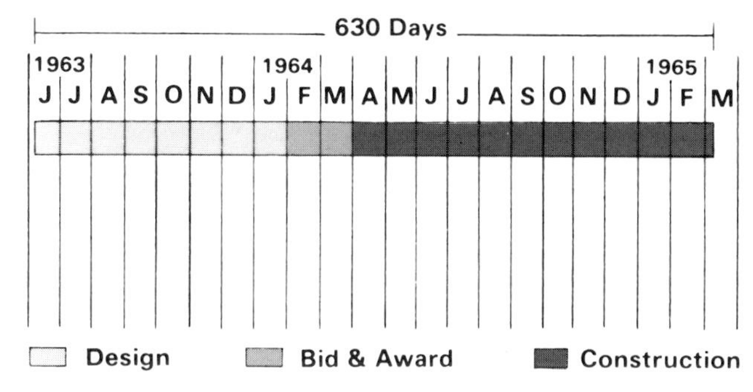

Total Cost..$14,200,000.00
Size: Baseball cap. - 51,500; football cap......58,000
Cost Per Seat (football)................................$244.83

This project was an important step in the growth of a rapidly developing city, developing both in urban growth and in economic stature. The timely completion of this project within the bond funds available, much more so than in most recreational or spectator facilities, was not only important to the owner, as it usually is to any owner, but represented a crucial impetus to development in the City of Atlanta.

This project illustrates a number of points relative to construction management that is utilized with a traditional, single-responsibility, lump-sum general contract for which the contractor is selected through competitive bidding. As is pointed out in Chapter 2 in a definition of construction management, there are many misconceptions regarding construction management and single-responsibility construction contracts. There is no less need for a construction management effort, and it has no less value to an owner, when a single-responsibility general contract is used than when multiple contracts are used. As the definition points out, there is little difference in the effort put forth preconstruction under this construction management contract plan than there is under the multiple-contract format. And even during the construction phase, there is little if any difference in the value construction management has to the owner in the two approaches. The only difference involves the manpower requirements of the construction phase.

This project also illustrates that the designer/construction manager still has many ways with which to deal with potential construction constraints under the single-award construction-contract format when phasing is not an available tool. Following the procedures for use of the Critical Path Method of scheduling that are discussed later in Chapter 9, during the preparation of the contract documents for this project, it was recognized that a major constraint would be the pile-driving operation. This was due to the condition that a good bit of grading and excavation was required and that if pile driving started at the normal time (i.e., after the building area excavation) and subsequently proceeded at a normal pace (i.e., one shift a day), that structural steel would arrive long before the foundation work was completed. In turn, it would be impossible to meet the scheduled twelve-month construction time. Consequently, an atypical construction sequencing plan was adopted that was the basis of the preliminary construction program management network issued with the bidding documents. This plan called for pile driving to start immediately after award of the construction contract and for piles to be driven through overfill in areas of cut and driven in such a way as to receive fill in areas requiring fill. Also, a three-shift pile-driving operation was scheduled. Thus, the plan for the first three or four weeks of the construction program, which were crucial to the completion date, was worked out in one feasible manner, in advance, by the designer/construction manager for the consideration and use of bidders. The successful bidder decided to accept this procedure, and that schedule was followed successfully during the early phases of the project.

Earlier in the design phase, the fabrication, delivery, and erection of the structural steel was recognized as the other major potential constraint in the construction schedule. Unable to deal with this condition by an early (phased) award of the structural frame due to the policy of the owner, this constraint was dealt with effectively through the medium of design. The structural steel frame was to start at the main concourse level of the stadium, all structural work below that level being cast-in-place concrete. All of the major structure above that level, with the exception of the precast concrete seating risers, was designed as steel. In this case, a tradeoff study, as a design phase construction management activity, was carried out by the designer/construction management team, and steel, in this instance, was selected on the basis of time, cost having been shown to be an even tradeoff. However, it had been seen that in designs for other similar multipurpose stadiums that were round or nearly round in configuration, that anywhere from four to twenty different structural bent designs had been employed to accommodate varying conditions of field lighting loads, seating configurations, etc. By studying the fabrication process that would be required for steel members of this size and type, it became clear that substantial savings in fabrication time and in erection learning time at the job site could be realized by working toward the most orderly and standardized structural frames possible. In turn, it developed that it was possible to have a single uniform bent design that was repeated eighty times in the project. Therefore, it is estimated that roughly sixty to ninety additional days (a combination of shop fabrication and field erection time) were saved through this design effort that, in turn, resulted from construction management information being available and utilized in the early design phase.

This project was completed for beneficial occupancy six days ahead of schedule. In this case, beneficial occupancy was very clearly understood in that a major-league baseball game was played, to a full house, on the day of beneficial occupancy.

The Finch-Heery joint venture team, which worked well together during this project, decided to stay together as a permanent joint subsidiary organization of the two parent companies, and, in turn, has handled a series of sports and other major public facility projects since the Atlanta Stadium. One shortcoming in the design of the Atlanta Stadium was the absence of moving grandstands in this dual-purpose stadium. The use of major elements of moving grandstands, as this same professional team later used in the Cincinnati Riverfront Stadium, is important to the seating arrangement for both baseball and football in a dual-purpose facility. It becomes more crucial to football. Consequently, in the Atlanta Stadium the seats between the 30-yard line for about the first fifteen rows are not equal in quality of sightline and juxtaposition to corresponding seats in a single-purpose football stadium or in a dual-purpose stadium gaining the benefit of moving grandstands. The omission of the moving grandstands in the Atlanta Stadium was a predesign budgeting decision based on funds available and the state of the art of the development of the moving grandstands at the time of the design of the Atlanta Stadium rather than being an omission for budget adherence during the design phase.

TICKETS

CASE HISTORY 4

ENGINEERING OFFICE BUILDING FOR LOCKHEED AIRCRAFT CORPORATION, MARIETTA, GEORGIA

This Heery & Heery design and construction management project illustrates a number of points in the utilization of an industrialized building system (School Construction Systems Development—SCSD) in a highly accelerated design and construction program: the fact that the SCSD system may be applied to building types other than schools, and the use of the technique of phasing the construction contract award with subsequent transfers into a general contract. The latter is discussed in more detail in Chapter 10.

This version of SCSD was the early Inland Steel Products competition-winning system that has subsequently evolved into a somewhat more flexible construction system and is more widely and competitively available throughout the United States, as discussed in Chapter 6.

Some of the significant options open to owners, architects, engineers, and construction managers through the use of industrialized building systems and industry-accepted performance specifications are illustrated here. The time chart shown in this case history illustrates that early in the design phase bidding and award for major components of the building's system took place. (In all of the time charts in these case histories, the "design" bar indicates all design activities, including preliminary design as well as contract document preparation along with any required design research and the in-progress owner's reviews.)

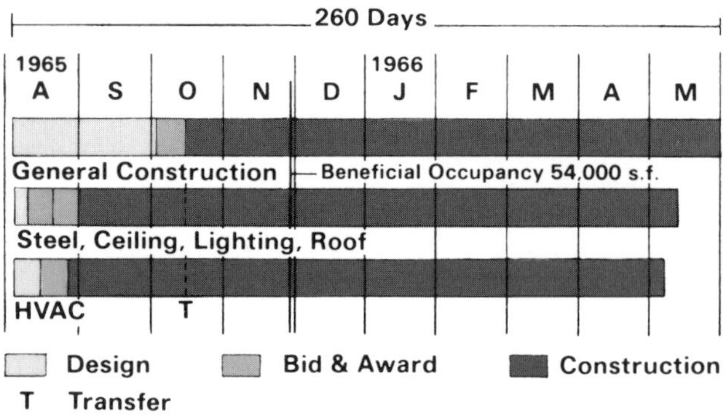

260 Days

1965 A S O N D 1966 J F M A M

General Construction — Beneficial Occupancy 54,000 s.f.

Steel, Ceiling, Lighting, Roof

HVAC T

Design — Bid & Award — Construction

T Transfer

Total Cost **$4,618,600.00**
Size .... **306,000 sq. ft.**
Cost Per Square Foot **$15.10**

In preparation for this project, the Lockheed Facilities Group, a highly professional group of managers headed by Facilities Director Emil Docekal and Facilities Manager Jim Denny, were well prepared with a definitive program, a selected site, and site data. In turn, in the predesign project analysis, as is discussed later in Chapter 5, with owner participation, it was determined that the SCSD system would be utilized. With this early decision and site/program information available, it was then possible to issue bidding documents for the SCSD structural frame, ceiling and lighting system, interior movable partitions, roof deck, roof insulation and roofing, fascia and fascia insulation, making an award of these components of the building fifteen days after the commencement of the design. This early award package amounted to approximately $2 million and included erection. It was possible to do this by only arriving at this point in the design process: adopting the perimeter "footprint" configuration of the building, selecting the bay size, accepting the 5-foot planning grid of the SCSD system, and selecting four different ceiling lighting conditions with allowed quantities of each (options were retained for quantity variations on a unit price basis). Thus, the architect had not limited himself to the exterior skin design of the building, the internal arrangement of the building, or the selection or location of major plumbing, underfloor electrical, equipment, or architectural finishes. It was possible, at that time, to receive two competitive bids on the above-described building system, it having since become possible to obtain much wider competition for corresponding components of the derivative SCSD system.

Nine calendar days after the award of the contract for the SCSD components described above, the complete heating, ventilating, and air conditioning system was awarded on a directly negotiated basis. At that time, there was such limited competition for the HVAC (heating, ventilating, and air conditioning) component of the SCSD system that the construction management decision was to negotiate for this major system in the building. As in the other components of SCSD, since that time much wider competition has become available.

Subsequent to the award of the above-described subsystems, all of which represented major potential constraints in the construction, a general contractor was selected through prequalification and competitive bidding on flat fee predicated on the acceptance of the schedule, the proposed method of contract transfers, time completion conditions such as those described in Chapter 8, and an agreement on the method for completing pricing for the job. Once the general contractor was selected, the remainder of the project was priced by material suppliers and subcontractors quoting jointly to the architect/construction manager and the selected contractor. Upon the completion of all pricing, the early awarded contracts, which included the SCSD systems discussed heretofore plus the electric switchgear, were transferred into the general contract and the price resulting from the sum of the early awards, the additional prices received, and the contractor's flat fee became the lump-sum price of a single-responsibility general contract. Thus once construction started, the owner had a "traditional" general contract which, in turn, proved to be of tremendous advantage in eliminating exposures to delay and change orders.

The photograph of Lockheed Facilities Director Emil Docekal and the author was taken on the day of beneficial occupancy of the first 54,000 square feet of the building. It had been the owner's requirement that 50,000 square feet be available six months after the commencement of design; 54,000 square feet were occupied two weeks in advance of that schedule, as is seen from the time chart.

The building design, while simple and straightforward, has proved to be a highly acceptable one except for the coating on the sheet metal work on the exterior. In the intervening years there have been substantial improvements in the coatings available for this type of installation. The total cost of the building did not suffer from the highly accelerated schedule, as the unit cost illustrates. While not significantly lower than corresponding unit cost for comparable office space, it was definitely in line, as the building incorporated high levels of lighting, multizone air conditioning, a great deal of underfloor electrical work, and movable partitions.

CASE HISTORY 5

HEERY & HEERY OFFICE BUILDING

This project had as its design program the more or less typical requirements of a multitenant commercial office building. Since the owner (Heery & Heery, Inc., Architects-Engineers) is consistently in the construction market, a negotiated contract was used. (See the discussion on group I and group II owners, related to an owner's posture for purchasing construction, in Chapter 3.)

As in most commercial projects, due to financing arrangements, it was not feasible to utilize phasing. A lump-sum general contract best suited the requirements of lender and owner, and it was necessary to have all financial arrangements made before any construction contract commitments could be made by the owner.

Therefore, the construction management plan was for a single-responsibility, negotiated, lump-sum general contract. Included in the contract price were quantities of tenant fit-up items such as interior partitions and doors, floor coverings, etc., with agreed-on unit prices for both additive and deductive price adjustments (maximum 25 percent deductive at negotiated unit price) for variations in quantities of the tenant work.

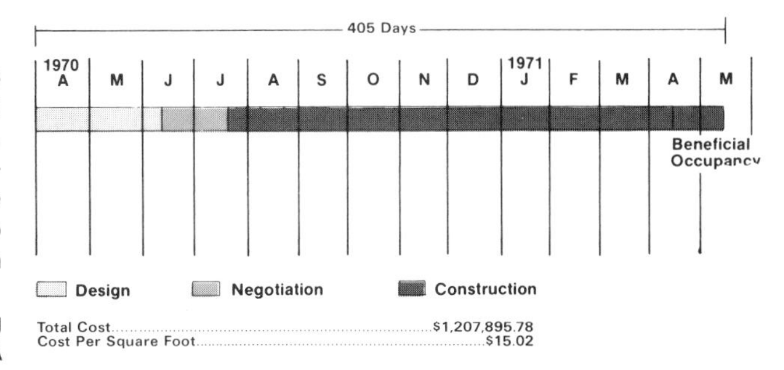

The total construction cost of the project was $1,207,895.78, which included all parking facilities, landscaping, and tenant work. On a square-foot basis, deducting $7 per square foot for covered and deck parking, the office space cost (with tenant work) was $15.02 per square foot of gross building area. Net rentable area as a percentage of gross is 88 percent.

CASE HISTORY 6

DELTA AIR LINES CORPORATE FACILITIES AT GREENBRIAR, ATLANTA, GEORGIA

This group of buildings is located on a rolling wooded site in suburban Atlanta. The architects masterplanned the site to receive several of the air lines' facilities including these, its systemwide computer center, and future general corporate offices.

The four-component Greenbriar complex illustrated here included the air lines' flight simulator facility, flight training center, communications and operations center, and regional reservations center.

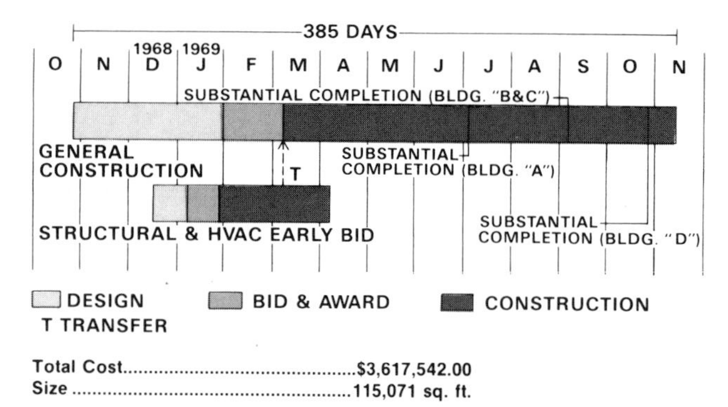

Total Cost..$3,617,542.00
Size ....115,071 sq. ft.

The project contains mechanical and electrical systems for maintenance of large quantities of computers, communications equipment, flight simulator, and other sophisticated equipment and functions.

Phased occupancy was a requirement to coincide with the air lines' vacating of corresponding leased quarters in downtown Atlanta and at the nearby airport as well as for simulator units delivery.

A major time constraint was off-site utility and site access work, two very common constraints that any architect/engineer or construction manager should always look into as early as feasible. In turn, the delayed site access, identified in predesign scheduling, caused the structural frame (selected as precast concrete) and the electrical switchgear to become secondary constraints. Consequently the management plan developed in the predesign project analysis (see Chapter 5) was basically as follows (see accompanying time chart):

It was feasible to complete all contract documents by the time off-site utility and site access work could start. Consequently, the advantages of a lengthened general contract bid time could be realized. The off-site utility and site access work could be accomplished in about forty calendar days; therefore, a forty-five-day bid and award phase was scheduled. However, bids for the structural frame and electrical switchgear were separately received, and those contracts were awarded, for early fabrication starts, during the first fifteen days of the aforementioned forty-five-day general contract bid/award period.

Both the general contract documents and the early awarded contracts for structural frame and electrical switchgear included provisions for transfer of contracts. Consequently, after bids were received for the general contract (Van Winkle Construction Company, low bidder), the early award work was transferred into the general contract. Thus, through this construction management plan, the owner had a single-responsibility contract once work at the jobsite commenced.

Heery & Heery were architects, engineers, and construction managers. Delta Air Lines' Facilities Department represented the owner.

LINK

CASE HISTORY 7

DELTA AIR LINES COMPUTER CENTER ATLANTA, GEORGIA

The Computer Center project for Delta Air Lines was undertaken in 1967 and completed to the point of beneficial occupancy during that calendar year. The client was represented by Mr. Hollis Harris, then Director of Facilities for Delta Air Lines. The size and cost figures shown with the time chart here include all building construction costs, exclusive of furniture but including all site and landscape work and the construction of a building to house the uninterruptable power system for the computer center. In the predesign project analysis (See Chapter 5), it was determined that the structural system best meeting the insurance, time, and cost requirements for this project would be a precast concrete frame. And in turn, the offsite casting, delivery, and erection of that frame was identified as a potential constraint and therefore dealt with by out-of-phase design, early bidding, and early award. However, it was determined that it would be in the best interest of this client to have as a part of the construction management plan the transfer of that contract for the precast into the general contract once the general contractor was selected. The general contract was awarded after competitive bidding by a selected list of bidders.

Three of the four photographs here illustrate the project as completed at the end of the first phase. The fourth photograph at lower right, shows the building after vertical expansion that was provided for in the original design.

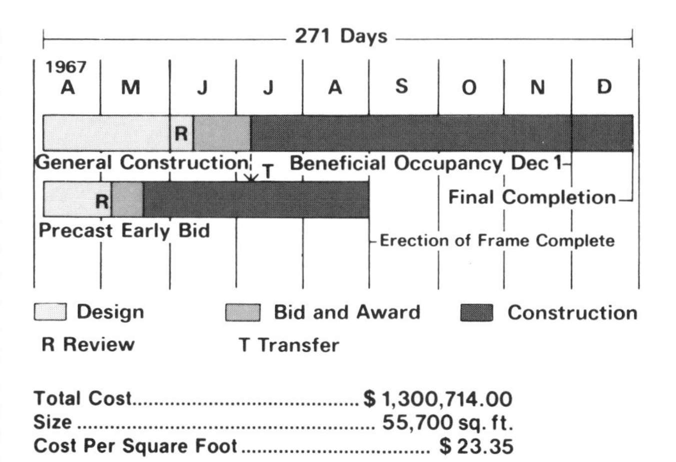

Total Cost.......... $ 1,300,714.00
Size .... 55,700 sq. ft.
Cost Per Square Foot .... $ 23.35

CASE HISTORY 8

SANFORD STADIUM
UNIVERSITY OF GEORGIA

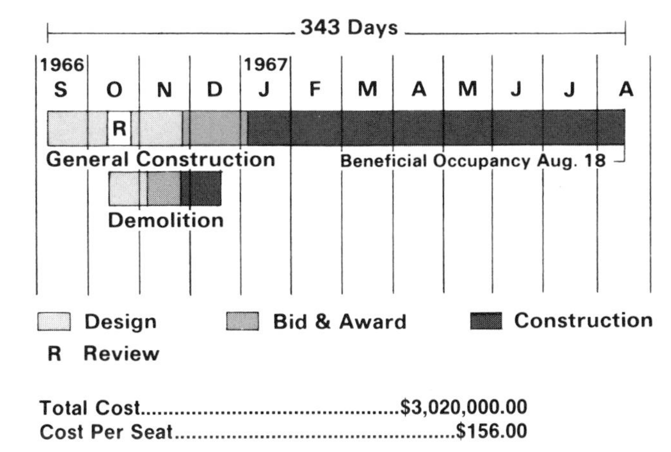

Total Cost..**$3,020,000.00**
Cost Per Seat..**$156.00**

The site for this project was that of an existing stadium with a capacity of about forty thousand seats. The older structure had taken advantage of a handsome natural bowl and both its lower-level riser slabs on grade and its playing field were in usable condition. All existing seating, rest rooms, press box, and the like were removed along with a good bit of seating that had been added through several expansions in a substandard manner.

The new structure was designed as cast-in-place concrete.

Heery & Heery provided architectural, engineering, and construction management services for the owner, and were authorized to proceed with the development of the design in September of one year with occupancy being required in September the following year. A further constraint was that none of the demolition could take place prior to the end of that fall's football season, which lasted until December 1 in that year's schedule.

Contract documents were completed, bids were obtained, and the contract was awarded prior to December 1. Consequently, since there was sufficient

time to allow for a general contract awarded in advance and since there were no construction constraint activities requiring more than two or three weeks preceding commencement of site work, there was no point in further phasing or separation of work into multiple contracts. This project illustrated how valuable Critical Path Method scheduling can be in making contract time extension rulings. This case is discussed in detail in Chapter 9.

Even though the design here called for many special conditions in concrete forming and almost all the work was exposed to effects of the weather, the very tight schedule was accomplished. All of the time-control actions and provisions that are discussed in Chapter 7 were brought to bear and found useful on this project. Likewise, the cost-control system described in Chapter 6 was utilized, the total project cost being considered exceptionally favorable for a project of this nature and size.

CASE HISTORY 9

THREE SCHOOLS FOR BROWARD COUNTY (FORT LAUDERDALE) FLORIDA

This project illustrates several techniques of design, scheduling, project management, and joint purchasing of industrialized building systems components. Design and construction management were provided by Heery & Heery with certain assisting services provided by associate consulting architect William G. Crawford, AIA. Joint systems purchasing was carried out jointly with Caudill Rowlett and Scott, architects and engineers for three projects of comparable program.

Major accomplishments in these projects were significant time savings within relatively low construction costs and a high degree of flexibility in very human spaces.

This school system, rapidly growing and facing the prospect of double sessions in the near future, had been requiring around three years for design and construction of a typical high school. (Actually, not a bad record when compared to national norms in urban areas of four to five years for senior high schools.) The aim was to cut the three years in half without escalating costs and while improving flexibility for a school system feeling its way into new teaching methods. The need for new approaches to design and construction management were perceived by the nationally prominent educator Dr. Benjamin C. Willis, who was then superintendent of the Broward County Schools. It was Dr. Willis who set the stage and provided the owner backing for the architects and construction managers to bring in new methods and systems, which were frequently controversial in the community but which successfully accomplished the stated goals. The basic program consisted of a 2,200-student senior high school (Piper High), a 1,000-student middle school (Crystal Lakes), and a 500-student-plus-kindergarten elementary school (Park Ridge).

Predesign project analysis was accomplished in roughly seventy-two hours by the Heery & Heery team working in temporary quarters in the locale. (See Chapter 5.) In these multidiscipline sessions it was determined that the School Construction Systems Development (SCSD) system would be applicable, and this was also the analysis of Caudill Rowlett and Scott for their three projects. The major potential constraints were identified as the award dates for systems components contracts, the working drawings for the high school, and a demucking operation that was required on the high school site. Another concern was placing all six projects on the market for general construction bids at the same time. In turn, it was the construction management plan that:

1. The two architect/engineer teams would complete schematic design on all six schools simultaneously, on a highly accelerated basis. (Accomplished in approximately two weeks.)

2. With the schematics approved for all six schools, the two firms would collaborate on performance specifications bidding documents for the full requirements of the six schools for the structural frame and roof deck, heating, ventilating and air conditioning, ceiling/lighting, and movable partitions subsystems of the SCSD system. (This was accomplished, including provisions for both steel and precast concrete bids on the frame—precast won—and allowance for differences in configurations and details of the two firms' projects.)

3. The subsystems' respective low bidders would be identified with letters of intent to allow for better clarity in the general construction contract documents, but awards would only be made as necessary.

Also, cancellation rights at a nominal cost time and right after the scheduled general contract bid time were reserved for the owner. In these ways, constraints in systems bidding and award were dealt with while minimizing the owner's financial exposure. Also, rights were reserved to transfer these proposals and early awarded contracts into the subsequent general contracts. (This was all accomplished; however, early bids for carpet were rejected and successfully rebid later, within budget, and first structural frame bids were rejected, due to successful contractor bondability, and rebid within allotted time.) Following are Heery & Heery's subsystem estimates and actual bid/award prices:

	Estimate	*Bid/Award*	*B/A per square foot*
High School:			
Structure	$ 478,913	$ 506,000	$ 2.25
HVAC	532,125	462,000	2.06
Ceiling/Lighting	236,118	210,000	.94
Mov. Partitions	101,724	103,000	.46
Middle School:			
Structure	$ 253,856	$ 240,725	$ 2.04
HVAC	282,063	303,548	2.57
Ceiling/Lighting	136,950	125,700	1.07
Mov. Partitions	35,280	30,697	.26
Elementary:			
Structure	$ 97,200	$ 65,639	$ 1.44
HVAC	108,000	105,816	2.32
Ceiling/Lighting	59,738	65,200	1.43
Mov. Partitions	8,558	12,646	.28
Total, all three H & H Projects	$2,330,525	$3,362,271	$17.12

4. The demucking work at the high school site would be bid and awarded early and work completed prior to the award of the general contract. Therefore it would not be necessary or desirable to transfer it into the

general contract. (This plan was followed and accomplished.)

5. The high school would be bid and awarded first to allow for its longer construction time. The middle and elementary schools would be spaced out in their bid dates as much as feasible to diminish impact on the local market. Early acceptable proposals and awarded contracts for subsystems would be transferred into respective general contracts at the time of the general contract awards. (All this was accomplished. The cost figures shown for each school include the subsystems, the general contract bid prices, and all change orders, thereby giving the final cost. Previous projects of this owner had not included equipment and site work in general contracts; however, all those items were included in the contracts and the listed unit costs.)

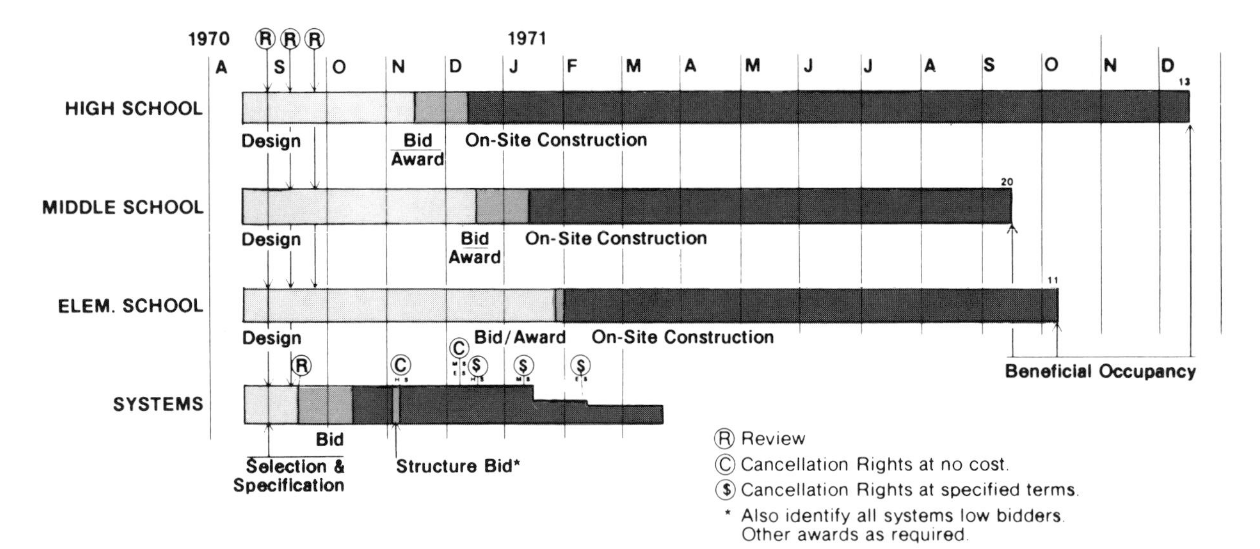

PIPER HIGH SCHOOL
Area in square feet 224,500
Construction cost exclusive of site work and equipment $4,688,461
Cost per square foot $20.88

CRYSTAL LAKE MIDDLE SCHOOL
Area in square feet 118,000
Construction cost exclusive of site work and equipment $2,264,819
Cost per square foot $19.19

PARK RIDGE ELEMENTARY SCHOOL
Area in square feet 45,650
Construction cost exclusive of site work and equipment $834,423
Cost per square foot $18.28

Piper High School

Crystal Lake Middle School

Park Ridge Elementary School

CASE HISTORY **10**

NATIONAL DISTRIBUTION CENTER AND DIVISION ADMINISTRATIVE OFFICES FOR LOCK AND HARDWARE DIVISION OF EATON CORPORATION, MONROE, NORTH CAROLINA

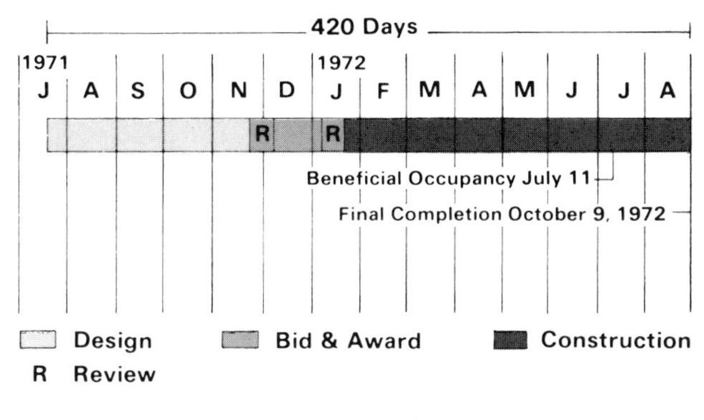

Total Cost..$1,637,313.53
Cost Per Square Foot....................................$13.53

This project consisted of basically two elements, a one-story office building which contained certain employee facilities, such as the cafeteria and the distribution warehouse. This example of a typical industrial distribution facility had as its constraints the structural steel frame and the electric switchgear, both quite common long-lead items in short-term, one-story industrial projects. The structural frame and electric switchgear were bid and awarded early and subsequently transferred into the general contract, which was also obtained by competitive bidding. All the time and cost control procedures discussed in Chapters 6 and 7 were utilized, as well as the critical path method of scheduling discussed in Chapter 9, techniques of phasing and transferring contracts, and other construction management procedures as described in Chapters 10, 11, and 12. Heery & Heery, Architects, Engineers & Construction Managers.

CASE HISTORY 11

WAGNER ELECTRIC PLANT
BOAZ, ALABAMA

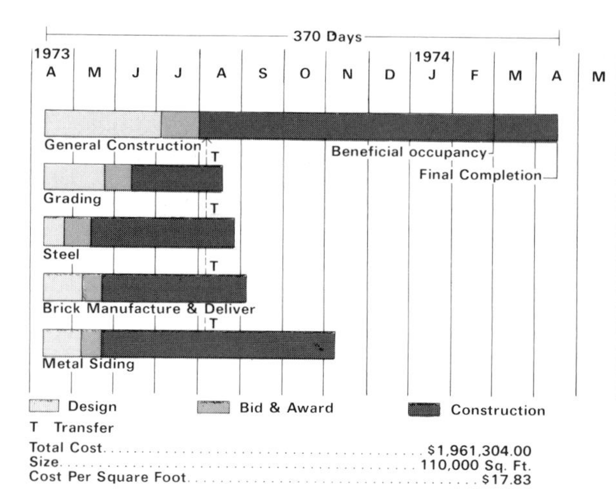

This typical light manufacturing facility was designed and built for the Wagner Electric Corporation of the Tung-Sol Group of the Studebaker-Worthington Corporation and is one of a series of such facilities designed throughout the southeastern United States for this owner by Heery & Heery, Architects and Engineers, who provided a combined design and construction management service. In this project the potential constraints that were identified during the predesign project analysis were general grading of a relatively large and rough site along with the rail spur construction; structural steel frame fabrication and delivery; metal siding delivery and erection, and purchase and delivery of brick. Contracts for these items of work were bid competitively and awarded early, being subsequently transferred into a competitively bid general contract. All elements of the time/cost control design and construction management system were applied throughout, as discussed in the above case histories. Total cost of construction including site work was $1,961,304. Gross floor area was 110,000 square feet, giving a unit cost of $17.83 per square foot.

CASE HISTORY 12

JUMBO JET EMPENNAGE MATE AND TRIM BUILDING

This Heery & Heery project was undertaken for the Lockheed Georgia Company, manufacturers of the C-5 Jumbo Jet transport, the largest airplane in the world. The aircraft was designed and manufactured at the company's large manufacturing facility in Marietta, Georgia, the plant having previously been used for the manufacturer of a number of other large aircraft. However, the high bays of the main existing assembly building did not provide sufficient height clearance for the tail assembly (empennage) of the C-5.

Consequently Lockheed developed a series of plans for handling this phase of the assembly of the aircraft and had tentatively settled on a facility utilizing a more complex approach that would have cost approximately $4.5 million. However, it was determined that this expenditure was excessive. In turn, Heery & Heery were engaged as architects, engineers, and construction managers for the needed facility. It was then necessary to also shorten design and construction time, as well as to reduce cost, in order to meet production schedules.

The proposed aircraft assembly operation involved over 11,000 separate structural, hydraulic, and electrical connections. Templates of the various components were furnished Heery & Heery by Lockheed's manufacturing-engineering group, and a space of approximately 100 feet in width by 200 feet in length and approximately 100 feet in height was developed in model form. With Lockheed staff, a simplified manufacturing layout was refined into the schematic design for the project. The design was based on a decision that the aircraft would be backed into position, with only the affected portions of the aircraft contained within the assembly building and with the aircraft fuselage stabilized on jack points for the mating operation.

This meant that a weather enclosure would have to be made around the fuselage between the wings and the empennage assembly. In turn, this generated the requirement for a hangar door of great precision with an adjustable aperture. At the time of completion, this door was the largest hangar door with adjustable aperture ever constructed or installed.

Construction constraints were identified as the rolling schedule for the 36WF300 steel columns and the lead time for ordering the 10-ton radio-operated underslung crane. Consequently early contract documents were prepared for the structural steel frame and the bridge crane; bids were received and contracts were awarded for these two portions of the project prior to completion of the contract documents for the remainder of the work. All the remaining work was bid in a single-responsibility lump-sum general contract.

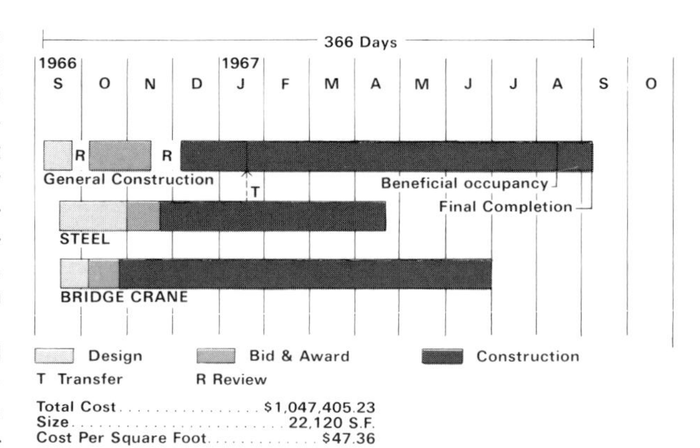

The low bid, including the transferred early award work and some additional work on the flight-line power-distribution system and central steam plant, was $1.25 million, representing a savings of over $3 million from the original proposal for the facility. First analysis of the design requirements was undertaken by the architects and engineers in February 1968, and the first aircraft was positioned in the building on September 4, 1969.

CASE HISTORY 13

TWO ELEMENTARY SCHOOLS CLARKE COUNTY, GEORGIA

In this case history two industrialized systems elementary schools were designed and built for the Clarke County, Georgia, Board of Education. They were the Fowler Drive and Barnett Shoals Road Schools. In February 1966, an existing shortage of classroom space was further aggravated with the loss by fire of an existing older elementary school. On the day following the fire, February 10, Heery & Heery was asked to undertake an effort to provide two new, air-conditioned elementary schools by fall 1966, just over six months away. Even though this very ambitious schedule was a necessity for the school board, it was determined that these two new schools must also represent an increased level of quality in both educational planning and construction. For the first time the school district would be incorporating carpeting, movable partitions, and flexible lighting and mechanical systems, as well as planning for team teaching and easy expansion. Utilizing the SCSD (School Construction System Development) industrialized building system, awarding early contracts for the structural frame component as well as ceiling/lighting and central mechanical equipment, along with all of the other methods of the time/cost control system, the projects were accomplished on schedule and within budget. The SCSD components represented $260,226 out of a total construction cost of $672,458 for each school. Construction included all built-in equipment, food service equipment, and basic site development at $15.89 per square foot.

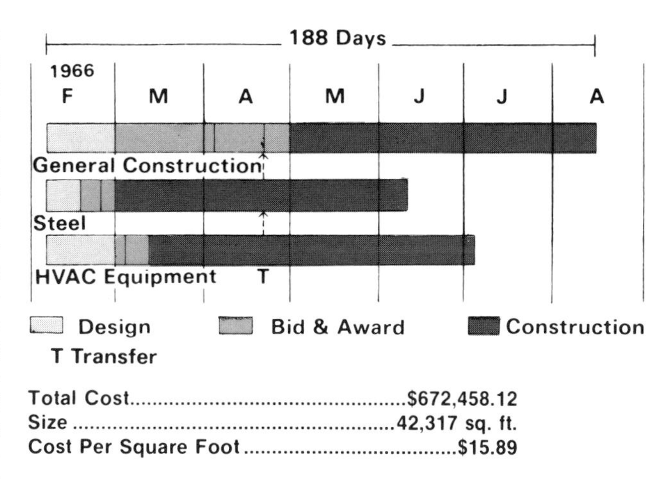

Total Cost....................$672,458.12
Size....................42,317 sq. ft.
Cost Per Square Foot....................$15.89

SIT AND READ

CASE HISTORY 14

GREATER CINCINNATI AIRPORT

The Greater Cincinnati Airport represented a major departure in airport terminal complex design. The project constituted the first effort in the airport/airline terminal field for Heery & Heery, Architects & Engineers, who served both the design and construction management roles under a single-responsibility contract with the Airport Authority. In this project, the Airport Authority, with Mr. Barry Craig as Director of Aviation, Robert Keefe as Director of Planning, and Mr. William Reeves as Project Manager, along with the major carriers (in particular American Airlines and Delta Airlines), as a group, identified for the architects/construction managers several major problems that had become most serious for the air transport industry: very rapidly growing construction costs that had reached a level of roughly $2 million to $3 million per passenger gate, design and construction times running from five to seven years, airport designs that were either difficult to expand or which only partially solved the problems of expansion, and lack of flexibility for the constantly changing operations that must take place inside the terminal buildings and in the surrounding support facilities. It was noted that most airport design projects in the past had been undertaken basically as an effort to design and build a civic monument of a static nature. The design direction that evolved here came about as a result of looking on the airport terminal complex as a system for handling passengers, baggage, aircraft, greeters, and all the other persons and operations associated with air travel. A systems approach to the design resulted in basically a "dry" and highly repetitive construction

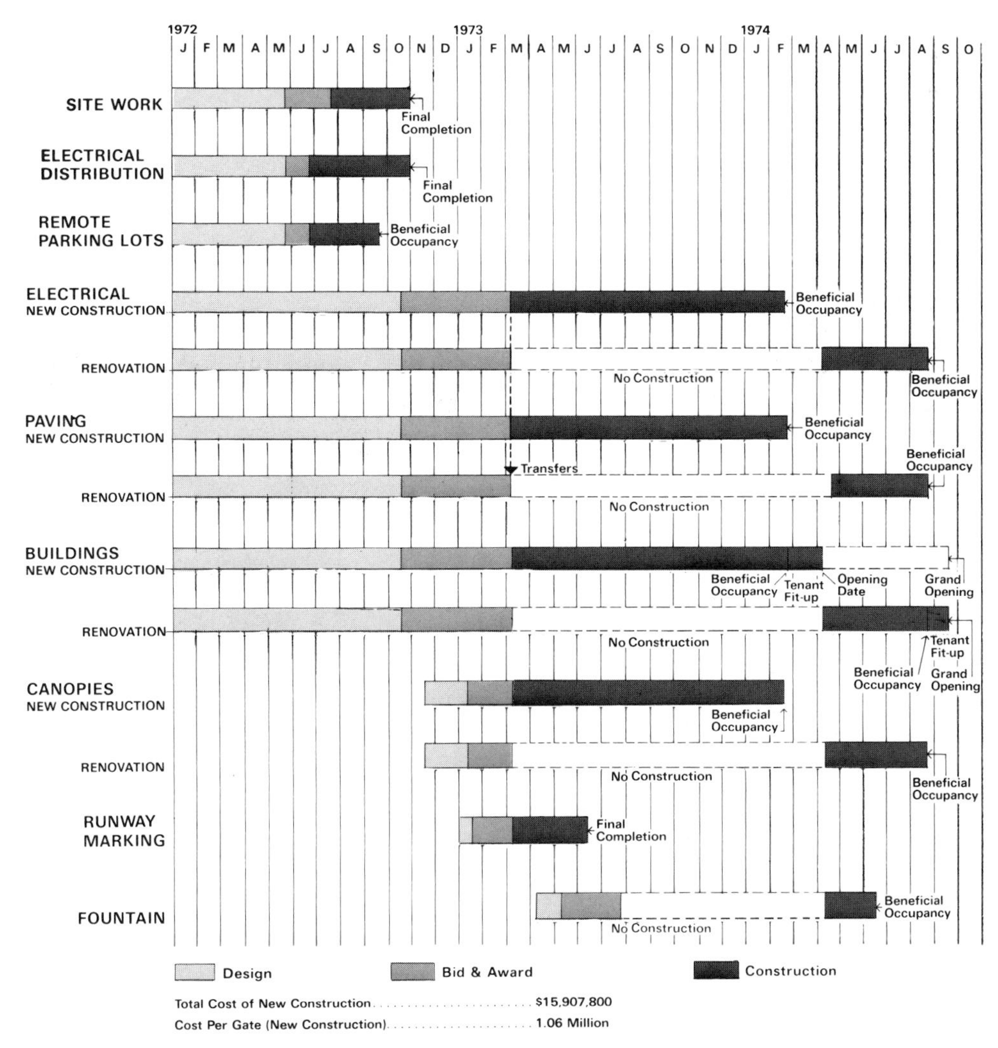

with a high degree of off-site fabrication. Due to the physical separation of much of the work and the necessity of dealing with potential constraints out of conventional phase, nine different contracts were awarded, only one of which was transferred into another contract. The overall terminal plan that was developed also had interesting aspects as to the passenger's convenience. For example, the baggage claim operation that is usually contained in the main terminal building was separated and placed on the parking-lot side of the roadway system. This and other aspects of the design allowed for the elimination of the expensive two-level road system frequently found in airports and thus made surface parking convenient. A major aspect of the expansion was the linear plan for the unit terminal complex. The overall result is a much less ostentatious complex that bespeaks its role in a transportation system, that was put into operation in a fraction of the time normally required, and that involved costs far below those usually associated with major airport terminal construction. This project also illustrates one significant variation of combining design and construction management. An early programming and budgeting contract was carried out prior to the main design and construction management contract, so that there was an agreed on schedule and budget for the architect/construction manager and his owner to include as a part of the main design and construction management agreement. Costs shown in chart prorate into the figure of $1,060,000 per passenger gate for the new terminals, including proration of all support facilities, parking, roadways, aprons, etc., but removing all costs related to the renovation of the existing building.

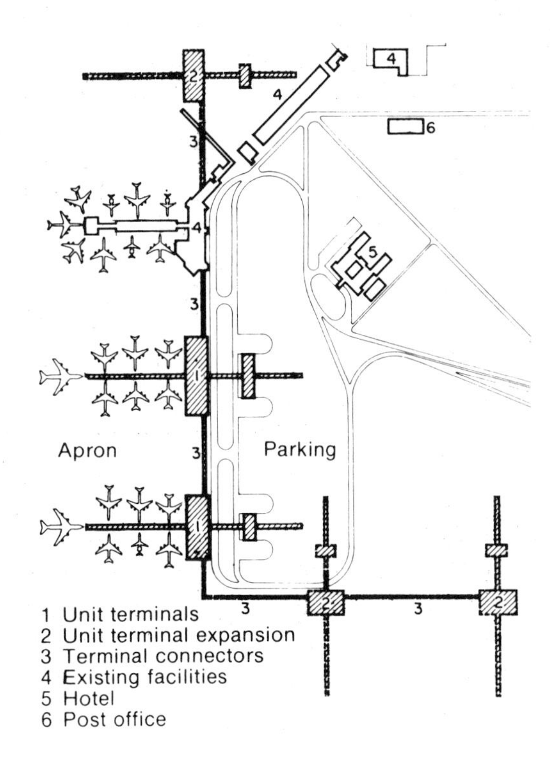

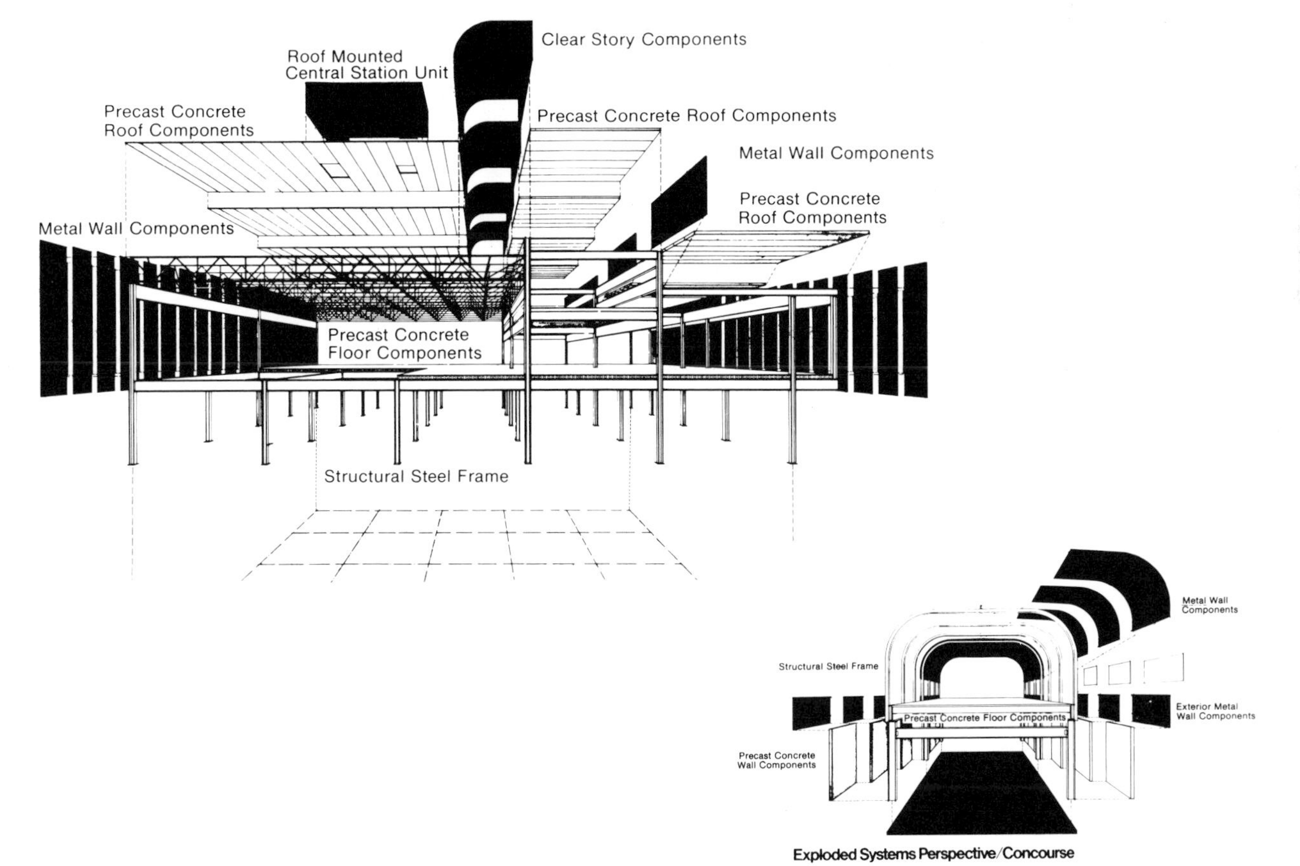

Exploded Systems Perspective/Concourse

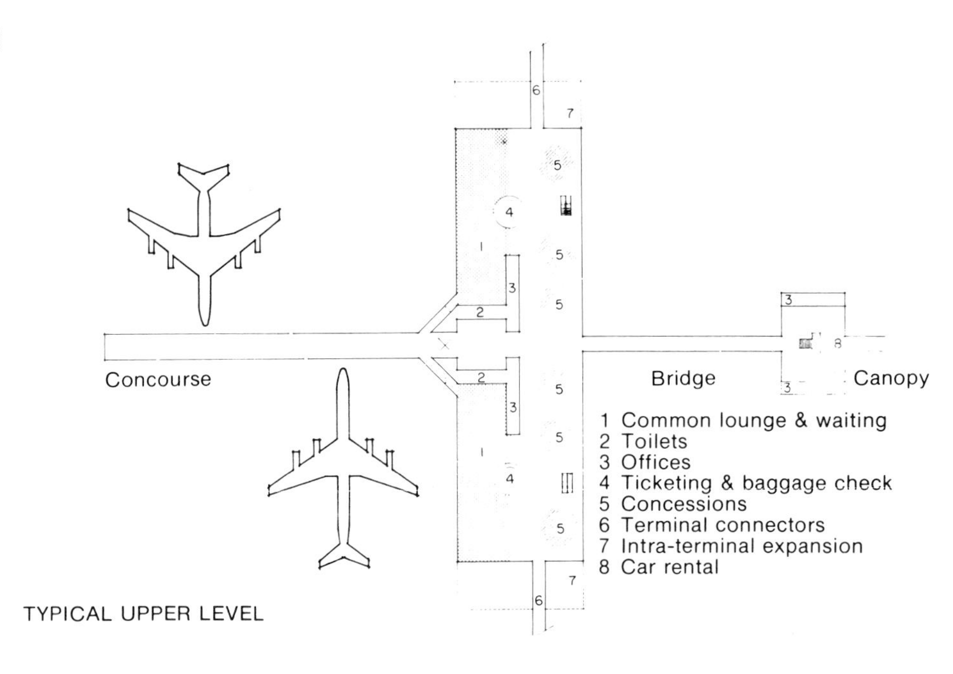
Concourse
Bridge
Canopy
1 Common lounge & waiting
2 Toilets
3 Offices
4 Ticketing & baggage check
5 Concessions
6 Terminal connectors
7 Intra-terminal expansion
8 Car rental
TYPICAL UPPER LEVEL

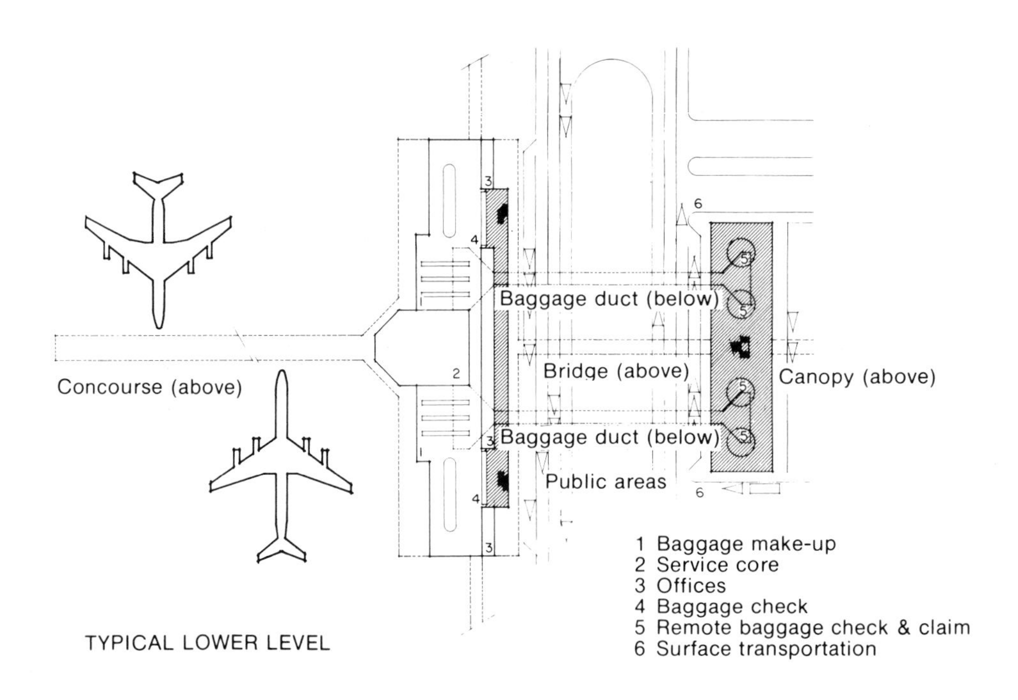
Baggage duct (below)
Concourse (above)
Bridge (above)
Canopy (above)
Baggage duct (below)
Public areas
1 Baggage make-up
2 Service core
3 Offices
4 Baggage check
5 Remote baggage check & claim
6 Surface transportation
TYPICAL LOWER LEVEL

DELTA
AMERICAN
TWA

Up to
DELTA
ALL OTHER FLIGHTS

AMERICAN
TWA

DELTA

Do Not
Enter

A
B

RIVER QUEEN
DINING ROOM
ALLEGHENY
NORTH CENTRAL
PIEDMONT

CHAPTER 2

A Management Approach to the Construction Program

Toward the latter part of the 1960s, the policy makers and managers of major building programs, for government and industry alike, began exhibiting an increased interest in a wide range of services, approaches, and systems that could be placed under the broad umbrella label of "construction program management" (In Canada, akin to "Project Management.")

The term construction program management is used here to indicate a broader set of management services related to the construction program. It is thus differentiated from the term "Construction Management" as the latter has become commonly used. Construction program management may incorporate construction management, plus a number of other professional services that can be productive in time and cost control of design and construction. It may also include certain other services that contribute to the functional quality and economic success of a project.

The raison d'etre for construction management is the control of time and cost in building construction.

While there is far from agreement on just what the correct definition of construction management should be, most professionals and firms operating in the field are providing a service somewhere between simply the direct management of the construction operation (See Chapter 12) and a somewhat more extensive set of services that incorporate the foregoing along with estimating and cost control, scheduling and information systems, time control and expediting, and bid/negotiation/purchasing management. It may, in some cases, provide one alternative to the traditional general contractor, though this should not be a predetermined conclusion.

Any one or more of the foregoing might reasonably be called a construction management service. All of them together would comprise a complete and typical construction management service to be provided concurrently with architectural and engineering services. See Figures 2-1 through 2-4, with accompanying notes, for various project management organizational plans.

Here may be an appropriate place to propose a more completely formed definition of "construction management" than has been heretofore accepted or in prevalent use.

Construction management is that group of management activities, over and above normal architectural and engineering services, related to a construction program—carried out during the predesign, design, and construction phases—that contributes to the control of time and cost in the construction of a new facility.

The professional construction manager, then, is the individual or firm who ties himself to an owner in a professional arrangement and applies the proper combination of management activities to a construction project to achieve time and cost control.

FIGURE 2-1

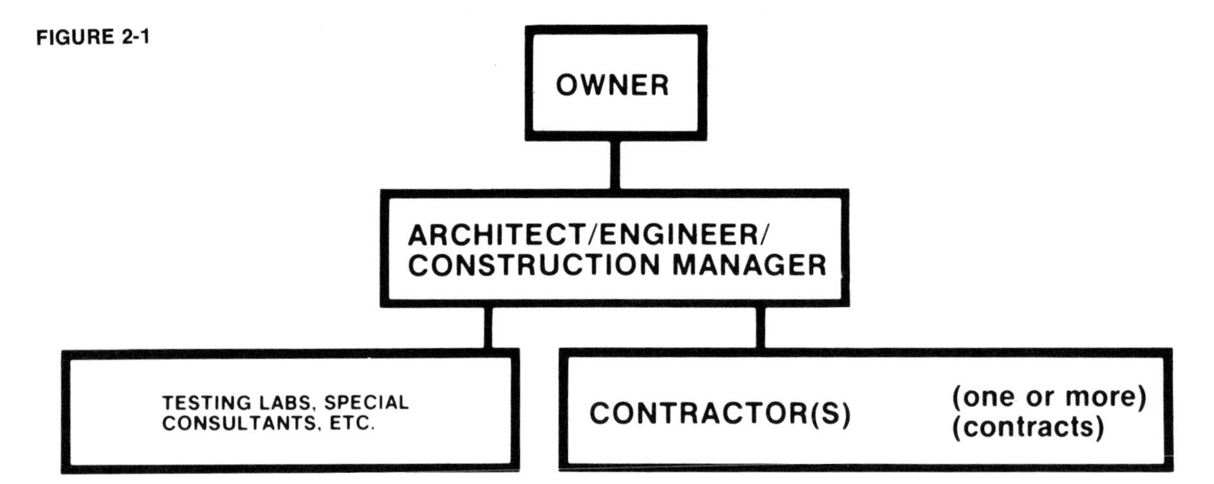

This chart illustrates the most simple and direct manner, along with probably the most economical approach, to an owner obtaining construction management services. Here, a combination, single responsibility, design and construction management service is provided by one firm having the capabilities of architecture, engineering, and construction management within its organization. While this configuration does not provide for certain advantages that the separate or "adversary" construction manager may offer to an owner, it may frequently offer other advantages that more than offset the absence of the foregoing arrangement. In many cases, this arrangement will be the best of the options available to an owner, assuming that the firm with appropriate capabilities is available to the owner for the project.

FIGURE 2-2

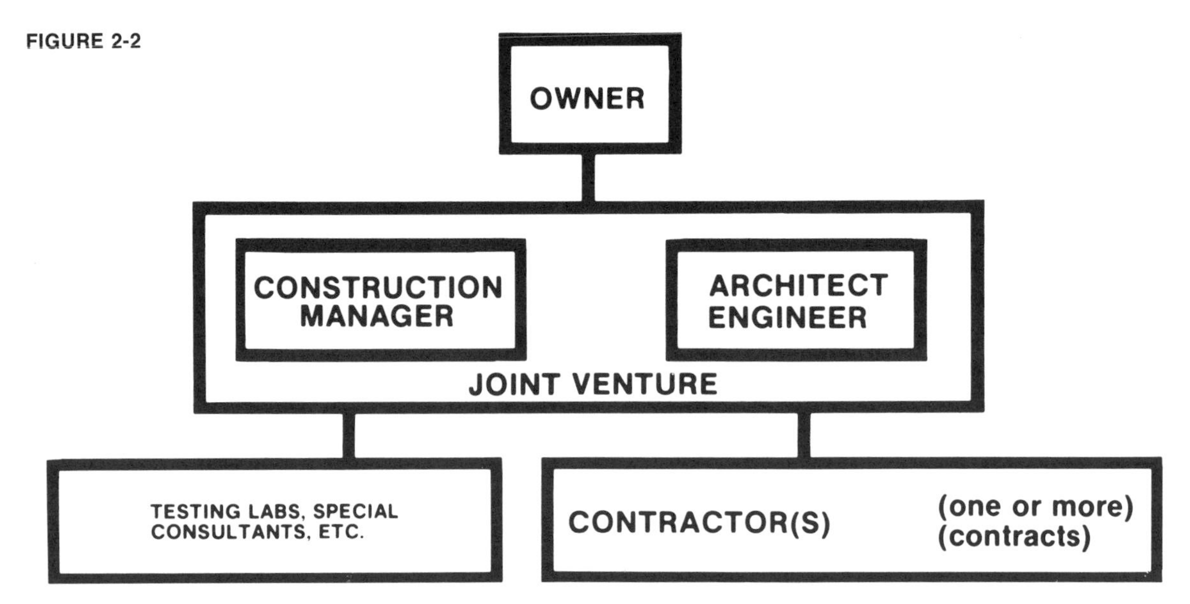

Basically the same configuration as illustrated in Figure 2-1, but with the design and construction management, single responsibility service provided by a joint venture. It is in the best interest of both the owner and the joint venture entity in this situation for the joint venture to have been voluntarily formed in advance of its offer of services to the owner, and for an agreement to be reached that provides a definitive workable arrangement between the construction manager and the architect/engineer. The advantages of this joint venture would be partially lost were there to be more than a single project manager representing the joint venture firm as the joint venture's primary contact with the owner throughout all phases of the design and construction program.

FIGURE 2-3

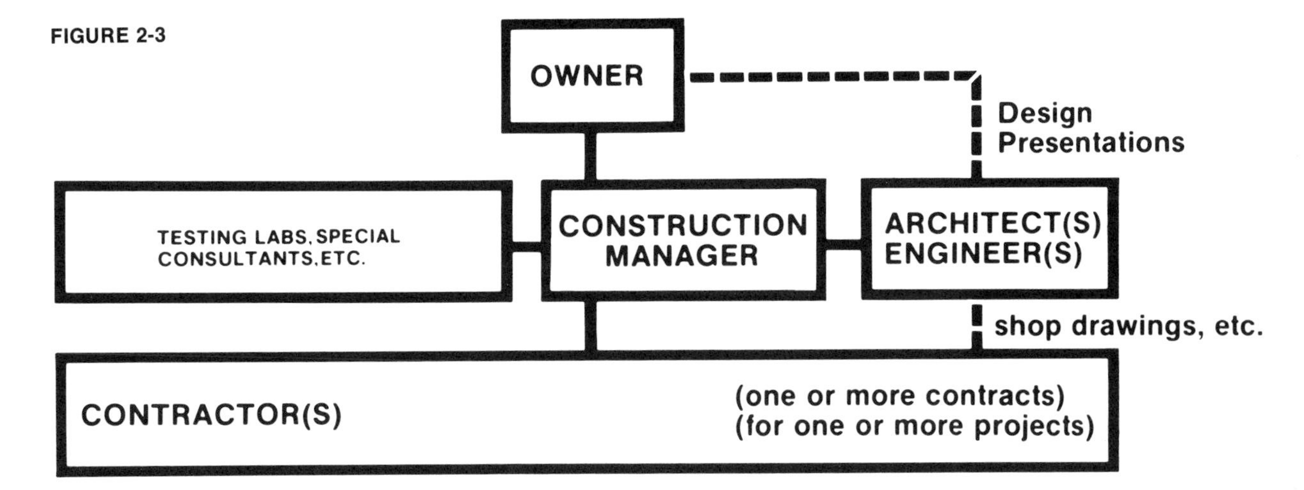

This illustrates a good and typical arrangement when the construction manager is a separate entity from the architect/engineer. This arrangement can have very nearly as many advantages to the owner, and in some cases more advantages, than those illustrated in Figures 2-1 and 2-2. Many projects wherein a construction manager separate from the architect/engineer is used may start out with some different organizational plan from that illustrated here, but if the construction manager is to carry out his functions properly a plan basically similar to this will emerge in most instances.

FIGURE 2-4

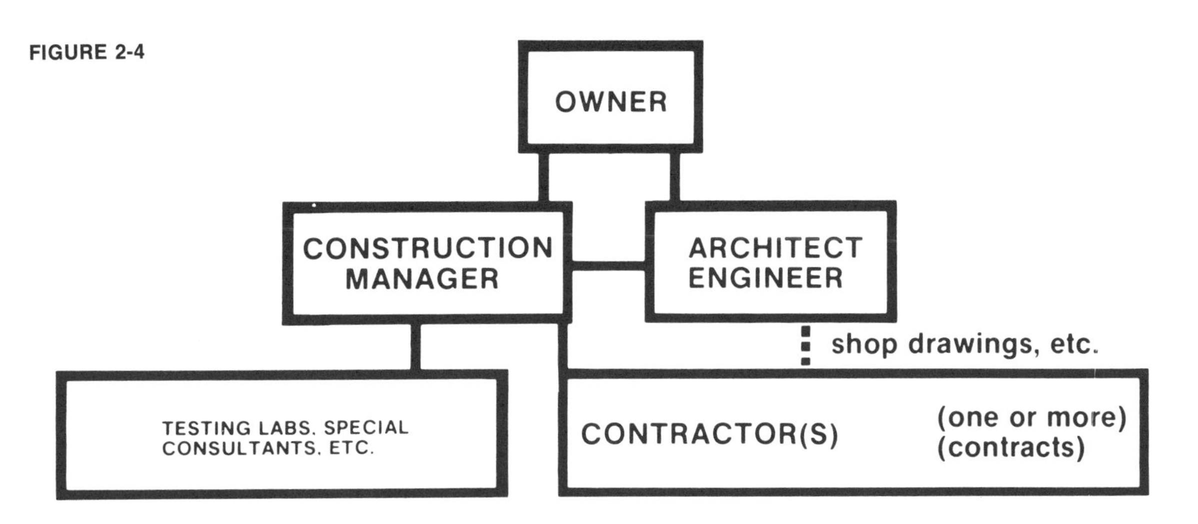

The relationship illustrated here between owner, construction manager, and architect is an arrangement often visualized in separate construction manager situations. However, this arrangement is not nearly as likely to be productive and efficient as is that arrangement illustrated in Figures 2-1, 2-2, or 2-3.

As a practical matter, professional construction management services do not include the direction of work within a contract or subcontract but are concerned with the planning for and manipulation of those contracts. If the construction management plan for a given project involves a series of eight separate contracts awarded in phases, it is neither the responsibility nor the proper function of the construction manager to undertake the contractual responsibilities of supervision, direction, and coordination of any of the work *within* any one of the eight contracts. This is an area of confusion among some who have not done their homework on the emerging professional service of construction management.

There is a school of thought, which is quite common today, that construction management is limited to the process in which the general contractor is eliminated and the construction manager administers a series of separate contracts under a professional fee arrangement with an owner. Of course this arrangement has become common and can certainly work well in some situations. *But* it tends to overlook a fundamental fact: the greatest savings in time and cost in a construction program can be achieved during the design phase. It is during the design phase that the quantity and quality of the building are established, the systems which will affect construction procedures are selected, and the start time for construction is determined. These activities almost invariably have more control over time and cost than management activities not initiated until bid/award or construction phases.

Another school of thought says that construction management means phasing of design and construction, overlapping the two functions to shorten total time. This "fast track" system has also been shown to work. But in many situations it has distinct shortcomings, including the fact that the owner cannot be assured of a final construction cost on a completed set of documents; that is, he cannot be certain just what he is going to get for his money or, for that matter, how much he will be required to spend.

Are these two approaches really construction management? Certainly they can be a part of it. But each approach is only one of many ways in which a project can be accomplished; at best, each offers only part of the control construction managers should provide.

In order to be most effective, the separate construction manager should always be involved during the design phase. Ideally, he should be employed *before* the design professionals. During the design period, the construction manager must monitor the activities of the design professionals to assure that the emerging design can be constructed within the budgeted time and cost. The construction manager will provide predesign programming and budgeting, cost analysis and value engineering, scheduling (along with guidance on materials and construction methods), and recommendations regarding the way in which the construction contract package should be arranged. The construction manager should generally determine the plan of construction. He will manage bidding and negotiations and handle contract awards.

Although there are any number of possible construction management plans for the construction phase, most will fall into one of four basic formats:

1. *A single-responsibility general contract,* obtained either through competitive bidding or direct negotiation, depending on the respective owner's posture and desires. It is in relation to this format, which might be referred to as the traditional approach, that the greatest misunderstanding exists. The idea that when one uses a single-responsibility general contract, construction management is not needed, or that when one uses construction management, the use of a single-responsibility general contract is precluded, comes about from a very narrow view of construction management. Or it may be that this view comes about from a lack of definition of the goals of construction management (i.e., optimum control of cost and time for the respective project). Assuming that competitive bidding is to be employed on a project, there are virtually no differences in the construction management activities that should be undertaken during the design and preconstruction phases whether there be a single-responsibility general contract or any number of other arrangements employing separate contracts for the construction work. Certainly, the design phase and other preconstruction activities of the construction manager are fully as important for this format as for any of the others, such as the three discussed below.

It is true that the scope of construction management services in connection with a single-responsibility general contract during the construction phase will be substantially smaller in terms of man-hours than the construction phase construction management services for any one of the three formats discussed below. However, they are hardly less valuable to the owner.

2. *Separate early contracts* awarded for portions of the project which are identified as potential constraints. Subsequently there would be the award of a general contract, and, by contract provisions in this and the early awarded contracts, one or more of the previously awarded contracts could be transferred into the general contract. The separate early awarded contracts might be for such things as the structural frame, electric switchgear, systems components, grading or foundations, etc. By the transfer procedure, as soon as feasible the owner has a single-responsibility general contract in effect for the project.

3. *Separate prime contracts* that remain separate throughout the project for statutory or phasing reasons. This approach is sometimes referred to as "fast track", as is 2 above. (However, there is no universally

accepted definition in usage for "fast track," and, in turn, a great deal of confusion seems to accompany the use of this catchy jargon.) Under this format, the separate prime contracts might include the major parts of the project such as mechanical work, electrical work, structural work, systems components, special equipment, site work, and general construction.

There is a variation to this format which is often found useful. The owner and construction manager may determine that the majority of the construction work should be included in a single "general" contract, but that one or more later separate contracts should be awarded for items such as carpeting, food service equipment, laboratory equipment, graphics and signing, or landscaping. Usually this plan is best during projects with long construction times which dictate that the items mentioned cannot be realistically priced during the general contract bidding period (because their purchase and installation occur late in the construction period).

4. *Direct management* of separate trade contracts and field crews. A separate trade contract is a contract for a portion as large as the mechanical system and as small as the tile work, glazing, painting, cabinet work, cleaning, or the like. Within this format it is not unusual to find thirty to forty separate trades on a project. Field crews (i.e., not separate contracts, but labor crews directly employed by the owner under the construction manager's supervision) might be used for such miscellaneous tasks as cleaning and unskilled labor work. In this case, and to some extent in approach 3, there is no actual general contractor. This approach, and approach 3 as outlined above, might frequently be employed in situations when the traditional owner-contractor relationship may not be in the owner's best interest.

Of course, construction management for a project may be obtained in other ways than by using a separate professional construction manager or a combined design and construction management contract. For example, the architect-contractor team approach might be said to automatically provide construction management to the owner. While this approach to a project is neither available to nor always in the best interest of all owners, it can be highly productive—with the right combination of architect and contractor and in the right set of circumstances—to the right owner. However, when this approach is employed, it should not be confused with construction management as a professional service provided on behalf of the owner.

Construction management provided through a combined design and construction management service by a single entity is, of course, a professional construction management service. And, assuming that a firm with this broader capability is available to the owner, this may frequently be the very best way for an owner to obtain construction management services. The fact that the construction management is not separate from the architect/engineer can easily have as many advantages to the owner as the separate construction management arrangement; it is a different but most viable and productive concept of construction management services.

To review, the function of the professional construction manager is to control time and cost on behalf of an owner during the construction process. In order to do this effectively, his efforts must begin concurrently with or before the design process. These efforts must continue throughout all phases of design and construction.

During the construction phase, the construction manager performs services which might be considered similar to the traditional architectural construction-phase services along with the work of the resident representatives, with all personnel representing the owner's interests merged into one team under the construction manager's direction. These include detailed observation of the appropriate liaison/coordination of architects and engineers, contract administration, payment control, and reporting to the owner. When there are two or more contractors on one site, the construction manager must coordinate the efforts of these separate contractors. But these services are only the basic services of the construction manager during construction. Other activities may include scheduling and critical path method system maintenance, the administration of owner or tenant move-in, coordination of off-site work, recommendations regarding recovery plans if contractors are behind schedule, preparation of maintenance manuals, and many other services depending on the needs, desires, and authorizations of the owner. These activities are the responsibility of the construction manager, adjusted appropriately for whichever construction management plan is in use and, of course, related to the construction manager's arrangement with the owner or client.

The most important point to remember, however, is that there are any number of ways to apply construction management. The one most applicable to a given project should be used regardless of whether or not it utilizes a single-responsibility construction contract or multiple contracts, regardless of whether or not it fits into a preconceived popular notion, and regardless of the fact that this will sometimes mean that fewer man-hours are charged to an owner than under alternative approaches. Once again:

Construction management is that group of management activities, over and above normal architectural and engineering services, related to a construction program—carried out during the predesign, design, and construction phases—that contributes to the control of time and cost in the construction of a new facility.

In both construction management and complete construction program management, demand has given rise to a number of new approaches. The need for some new approaches was evident to many. Arising from reasons ranging from frustrations with seeming failures of previous methods to needs for developing approaches for the best means of acquiring a new hardware in the construction industry, some extensive changes in the construction industry and in the professional service firms and agencies that deal with it began taking place in the late sixties and early seventies.

It is not hard to list some of the evidence that change had to come and will surely continue:

1. Constantly rising cost of the end product, an uncontrolled escalation that compares so unfavorably with the mass production cost control of other segments of the economy
2. Emerging new industrialized building systems, along with rapidly expanding roles of the mobile home, modular home, and preengineered building industries
3. Interim financing costs that were at substantially higher levels in the late sixties and early seventies
4. The fact that, in many construction contracts, the contractor is not obliged to accept the risk for time of completion and final cost
5. Significant changes in approaches and a serious desire for better construction program controls by many governmental agencies and officials, all too often accompanied by ever-increasing bureaucratic tangle
6. Labor problems, often seemingly insoluble, ranging from unrealistic demands, to low productivity, to illogical jurisdictional disputes
7. Design and construction time spans running so long in some cases that new facilities are obsolete before they are completed
8. The emergence of gigantic new multiple-project complexes that stagger the imagination, the expansions of which, in many cases, have no end in sight
9. An affinity by many architects and architectural critics for architectural designs that do not relate to the needs of our society, that ignore the functional requirements of the users, that are born of expensive and often archaic construction methods, and that can, at best, be classified as visually interesting medieval buildings
10. An environmental problem that has reached horrible proportions and upon which planning, design, engineering, and construction management have direct impacts
11. Overlying all, a mountainous backlog of need for housing, schools, health facilities, industrial plants, and all manner of urban developments and redevelopments

A new discipline called construction program management or construction management can hardly be expected to be the panacea for all of the foregoing. However, as the new profession is evolving in its various forms there would appear to be not only earnest attempts to deal with some of the foregoing problems and challenges but some significant progress being made in dealing with some of the eleven points listed above.

Still, some substantial number of failures will no doubt have occurred by the time construction management, as a clearly defined profession or service, has been more fully formulated. The magnitude of some of these failures may determine the rate of growth of construction management as a separate profession or service and, to some extent, its form. Reasons for these failures will probably include the following:

1. A readiness in some projects to unnecessarily abandon the owner's competitive position for purchasing construction. This attitude stems from the belief that—in projects that are to be highly accelerated—competitive purchasing would inevitably slow things down.
2. Failure on the part of some managers to place responsibility, with controls/incentives, on contractors, suppliers, and manufacturers for coordination, cost control, and early completion.
3. Proposals to use multiple contracts (phasing) when these procedures (which require more construction management personnel and involve a fee) may not be in the best interests of the owner for a given project.
4. Hard selling of unneeded or redundant services in some cases; in other words, an "oversell" of some construction management services.
5. Unnecessarily complicated project management staffing and organizational plans, accompanied by an overload of construction management personnel charged to the owner on a time and multiple of direct payroll expense basis. This is not to say that time-charge personnel is not an acceptable arrangement for construction management personnel. It is not only the most equitable for the owner and the professional service firm but probably the only practical approach for full-time and field personnel. But it is the unnecessary overstaffing with this type personnel, plus complicated organization plans, that often leave responsibilities unclear and too many parties reporting to the owner.

What, then, are the viable areas of service for construction program management, and what are going to be the real value benefits to the building owner or user? The services, in general, appear to divide into two categories:

A. Construction and contract management

B. Programming and planning related services

But there is a great deal of overlap, up to 100 percent in some cases, the latter probably being the best among possibilities. In any case, it may be that the profession or professional service of construction program management will never form into a stereotype service, which might well be one of its greatest attributes.

The following is a list of most of the major areas of construction program management activities. An asterisk or parenthetical note indicates that the respective service is discussed in more detail later in this or another chapter.

1. Economic feasibility studies and financing/funding assistance
2. Programming of facility requirements*
3. Architect and engineer selection and the consulting with and coordination of their work when construction management services are provided by other than the project architect/engineer*
4. Consulting in connection with industrialized building systems*
5. Cost control, estimating, and "value engineering"*
6. Consulting for contract document general and special conditions ("boiler plate" guide specifications)
7. Bid and negotiation management (see Chapter 11)
8. Scheduling and owner information systems (see Chapter 9)
9. Consulting on time-control procedures and methods*
10. Construction management (varying from the services discussed in Chapter 12 to the more complete service in the construction management services—which would probably include 5, 7, and 8 above—referred to in the first part of this chapter)
11. Complete construction program management services*

Even the following are quite legitimate activities of construction program management by either private firms or in-house staffs, assuming the respective capability exists:

12. Move-in planning and coordination
13. Facility operations and maintenance
14. Location services (Frequently utilized by industrial clients, this service would usually include site selection, tax and labor studies, transportation studies, and coordination of feasibility studies.)
15. Facility leasing program management
16. Operational planning
17. Industrial engineering direction and coordination

The following are more detailed discussions of those services listed above with asterisks:

Complete Construction Program Management Services (11 above)

The complete management of a corporation's, institution's, or governmental agency's construction program. This would normally include all services listed above in 1 through 9 and might well include any or all of services 11 through 17, depending on the owner.

An owner may have these services performed in-house (for the most part), or purchased entirely by contract for a single project, multiple projects, or on an annual basis for all projects.

Programming of Facility Requirements (2 above)

Frequently this service or activity is referred to simply as "programming," particularly within the architectural profession. However, this term can be confused with two other uses of the word: programming as in computer sciences or programming in the sense of programming the full construction program, from feasibility through program of requirement development and scheduling. But for the sake of brevity, the shorter term "programming" will be used below to mean programming of facility requirements. In the health facility field, a phase of programming is called "planning."

This service consists of the development of the owner's or user's facility requirements and translation into documentation so that it may be used for funding projections, site selection, design, construction management planning, and operational planning. Programming may also include research for and development of design criteria. Examples: development of the performance requirements of a closed circuit television system for a teaching hospital, development of performance requirements for the air conditioning system for an enclosed stadium, developing design or selection criteria for a movable partitioning system for an open-plan school, or development of performance criteria for an airport baggage handling system.

There is, of course, a limit to how far a programming effort should go in the predesign stage of a given project—a point of diminishing return. But it is the rare project that can get off to a good start without an adequate "program," "program of requirements," "design program," or whatever it might be called.

Owners have found, in an increasing number of instances, that a truly capable in-house staff for programming, sufficient in size to handle varying work loads of projects, cannot and should not be maintained.

For an owner's in-house facilities group to do the programming job by asking department heads to list their various requirements not only provides inadequate functional and performance criteria to the designers but invites requests for unneeded space or facilities. This approach usually results in a program couched in terms of so many square feet of office and manufacturing areas or so many classrooms, and it does not provide the designer with the operational insight he needs or a desirable degree of design latitude.

A good program will always contain a thorough narrative section describing the operation, functional goals, environmental needs and problems, economic goals, and philosophy of the facility need. It will then fully list all requirements in such a way as to provide the designer with maximum latitude while still tying down the essential requirements. It should include the results of research efforts, preferably in the form of design criteria. If the programmer is someone other than the designer, the program should place the responsibility on the designer for a "complete facility for the purpose for which it is intended."

One big dividend that the owner/user always hopes for but seldom gets from a programming effort is an overall analysis and trade-off study of his needs that will reduce the total size or cost of his project. The latter is admittedly a most difficult task to perform but one that will go far toward justifying the facility programming effort.

Programmers must have the opportunity to familiarize themselves firsthand with the nature of the operations to be housed and to interview users in depth. This is essential to a successful programming effort.

In more complex projects, such as health facilities, the construction manager may become the coordinator for a group of specialized consultants for programming and "planning," providing cost analyses in the predesign phase.

Architect/Engineer Selection and A/E Contract Management (3 above)

It is important to preface the discussion of this construction program management service with two statements:

1. Obviously, this service will not exist for the construction manager when the architectural and engineering services are being provided by the same firm that is providing the construction management services. And it is this situation—i.e., the combined design and construction management contract—that would appear to offer the owner the best of its options for the typical construction program. This is true whether or not the owner has an in-house facilities staff. The assumption is, of course, that the selected firm will have both the needed design and the needed construction management capabilities. It is true that frequently, in the past, many architects have not demonstrated a capability or a desire to provide effective time- and cost-control construction management. But there are certainly many other architectural and engineering firms in the United States who do offer these services and who have a good record of achievement in this area. When such a firm can be selected by the owner for the combined design and construction management contract, this is usually in the owner's best interest because

- It is simpler
- It should be less expensive to the owner
- It solves programming/design/construction management interfaces

Along some of the halls of Washington that are most steeped in bureaucracy, the combining of design and construction management service is thought to constitute a conflict of interests. Nothing could be further from the fact. While there are pros and cons for separation and combination of design and construction management services, in most typical situations the arguments for combination will at least offset the arguments for separation.

2. In the many situations when it is desirable or necessary to have a separate construction manager, this service—i.e., the selection of the architect/engineer and the negotiations and management of the architect/engineers contract with the owner—will be the most crucial act of the construction manager. At the same time, it can provide the construction manager with his most effective tool for managing the construction program.

If the construction manager who is separate from the architect/engineer is going to be effective, it is highly desirable that he be involved in the selection of the architect/engineer. At a minimum, he should be able to help prepare the agreement between the owner and the architect/engineer and to brief the architect/engineer before the execution of the agreement.

Also, in the case of the separate construction manager, the entire program will go more smoothly and produce better results if a complete program, budget, and schedule can be presented to the architect/engineer prior to completion of negotiations between the architect/engineer and the owner. These documents should then be made a part of the architect/engineer's contract with the owner. In these negotiations and project orientation sessions, the construction manager might be assisting or representing the owner.

Another important tool for the construction manager is his role in the owner's approvals of the various submissions of the architect/engineer. While the owner will almost certainly have the final word, the manager providing a complete construction program service might well be engaged to carry out the major part of technical reviews for the owner and be in a position to report on budget adherence as well as program and schedule adherences preparatory to the owner's approval at each stage.

Consulting in connection with Industrialized Building Systems (4 above)

The most effective use of industrialized building systems, such as SCSD and various modular units, will occur when the project architect, from his own experience, sees that an industrialized building system is the right solution for his project and then utilizes a system sympathetically and creatively. When this happens, a systems consultant will only be in the way.

However, because some architects have not had the opportunity to use systems or are not oriented toward their utilization, or because of instances of a desire on the part of an owner to investigate the use of, to utilize, or to develop systems for multiple projects that are to have individual project architects, the services of a knowledgeable consultant in the field can be most productive.

For these services to be effective, in addition to the technical services involved, the firm or individual providing this particular construction program management service should be put in a position to

1. Aid in the selection of, contract preparation for, and orientation of the architect(s)
2. Participate with the owner in design approvals
3. Prepare actual or specimen contract documents for bidding and award of system components contracts.
4. Coordinate with and advise the architect(s) on the technical and contractual matters relating to the use of the systems, including construction phase services

Cost Control, Estimating, and "Value Engineering" (5 above)

As has been mentioned before, the control of cost, along with control of time (a form of cost), is the underlying reason for construction management. This would include cost in all of its short- and long-term aspects. Included are specific tasks of estimating, analysis, continuous cost-control activities in conjunction with design, bid/negotiation/purchase management, and value engineering.

In the high-rise office building field some owners have made a practice of employing construction cost consultants for many years. Sometimes, unfortunately, the result has been that the "consultant" dealt the building design a series of devastating blows and then passed out of the picture, eschewing any continuing responsibility.

The most effective cost-control program will almost always be that carried out as an integral part of a combined design and construction management effort. The next equivalent, in the case of a separate construction management service, will be the same approach integrated as well as possible with the architect/engineer's design work. In both cases, the cost-control system that is a component of the time/cost control system (and is discussed in Chapter 7) would be one effective approach to use.

There will be many instances when a consultant or construction manager should be brought in by the owner or the architect/engineer to assist in cost control. When this is to be the case, the cost consultant must be brought in early. The firm or individuals being so employed should be put in a position to carry out, in close coordination with designers and project managers, all the functions described for the estimators and cost controllers in Chapter 6.

Value engineering is a popularized expression for professional, detailed analysis of a project or components of a project for construction cost reductions, corrections, or optimizations. The technique, in simplified terms, is that of carrying out an analysis on a component-by-component basis to study or develop alternative materials and methods that provide equivalent or acceptable results at lower costs. Value engineering has been developed into somewhat of a specialty in itself. The reader is referred to the work by Alphonse J. Dell'Isola.[1]

Consulting on Time-Control Procedures and Methods (9 above)

The ability to shorten design and construction time significantly within the budget and in so doing to set end dates dependably is the essence of construction management.

Time-control procedures and project expediting are integral parts of the construction manager's task whether he be a part of an architect/engineer organization, a part of a separate construction management firm, or a member of the owner's organization. Time-control procedures and recommendations that are a part of the time/cost control system are covered in Chapters 5, 8, 9, and 10. These or corresponding methods are the kinds of services that should be provided by the manager.

On the other hand, there can be situations when less than complete construction management services are needed by an owner or an architect/engineer firm. Then consultation and assistance limited to time control could be productive.

Under these circumstances, the time-control consultant will certainly need to be in on a predesign project analysis and preferably should carry out a precontract briefing for the architect/engineer. He will be needed throughout the project, specifically for the original schedule and management plant development, for "boiler plate" writing, and for consultation during the construction phase.

An important thing for the reader, particularly the owner, to understand is that time-control procedures initiated after design has commenced will not be as successful as if they had been initiated at the beginning of the design. Time-control procedures initiated after the first bid or construction contract package has been issued will be severely hampered. Time-control procedures or efforts to expedite a project undertaken after the work has begun at the site—without adequate provisions for time control in the contract documents—will be very expensive and probably ineffective for the most part.

[1]Alphonse J. Dell'Isola, *Value Engineering in Construction,* Construction Publishing Co., Inc., 1973.

ABOVE: Dusk view from across the Ohio River, looking toward the stadium and downtown Cincinnati. RIGHT: View of Riverfront Stadium looking southward toward Kentucky side of the Ohio River, photographed at the beginning of the 1971 World Series. [Heery & Heery—Finch, Alexander, Barnes, Rothschild & Paschal, associated architects and engineers (practicing in Ohio as Heery & Heery—Alexander & Rothschild).]

Sales, Training and Distribution Facility, Atlanta, Georgia. [Heery & Heery, Architects and Engineers (Model).]

Hotel in Life of Georgia Center. (Heery & Heery, Architects and Engineers.)

FOOD SCIENCE LABORATORY FOR THE UNIVERSITY OF GEORGIA

Food Science Laboratory for the University of Georgia at the University Experiment Station in Griffin, Georgia. (Heery & Heery, architects and engineers.)

BECTON-DICKINSON RESEARCH CENTER

Research Triangle, North Carolina. (Architects, Engineers, Landscape architects, Interiors, Graphics, Construction management: Heery & Heery.)

Architectural model of high-rise, multi-use, urban complex in Atlanta, Georgia. Heery & Heery, Architects, Engineers, and Planners, prepared the illustrated master plan and hotel in left foreground of photograph at left and in center background of photograph below. Lamberson, Plunkett, Shirley & Wooddall, Architects, designed the insurance company home office building seen in right foreground in photograph at left and to extreme left in photograph below.

CHAPTER 3

The Owner

This chapter discusses two general areas pertaining to the owner: First, it discusses the evaluation of the basic characteristics of owners for construction purchasing posture. Second, there are discussions of and recommendations on in-house facilities or owner's construction management liaison staffs of corporate, governmental, and institutional clients.

EVALUATING THE BASIC CHARACTERISTICS OF OWNERS FOR CONSTRUCTION PURCHASING POSTURE

In the preface of this book, three types of projects were eliminated from treatment herein. They are single-family residences, the industrial process plant (such as refineries and paper pulp mills in which the cost is predominantly that of process equipment), and heavy engineering projects (such as highways, dams, etc.). This leaves the central building construction market for commercial buildings; industrial, and other corporate structures; transportation, education, health, and research facilities; multifamily housing; and recreational, public assembly, institutional, and governmental projects.

The basic evaluation with regard to the owner that needs to be made in each case is the owner's posture for purchasing construction. In other words, what are the owner's potential problems and opportunities in buying construction?

A basic purpose of the evaluation is to determine whether a given owner should obtain competitive bids for construction contracts or whether he should consider direct negotiation.[2] There can be distinct advantages and drawbacks to both approaches, varying with the owner's posture. Time is not an important consideration in selecting between competitive bidding and negotiation, contrary to popular belief.

The important consideration in selecting between the two methods is the degree to which contractors and other elements in the construction industry are likely to be responsive to the owner because of the prospect of repeat business through direct negotiation.

To make this determination, the owner should be evaluated for characteristics that place him into one of the two very broad groups discussed below.

Group I Owners

This group includes those owners to whom contractors or other elements in the construction industry are not likely to be responsive because it is unlikely they can obtain repeat business through direct negotiation either because (1) owner is unlikely to have repeat projects in their area

[2] Including the contractor/architect team approach.

from the respective owner, (2) they are unable to identify individuals within the owner's organization who would directly control construction contract awards based on recent performance, or (3) the owner is prevented by law from negotiating directly for construction. This group would include, in most instances, the following categories of owners:

1. Very large corporations.

2. Other business owners who are not consistently in the construction market within a given geographic area.

3. Institutions that are not consistently in the construction market or that are prevented by policy or indenture from direct negotiation.

4. Federal, state, and local governments, agencies, and authorities who are prevented by law from direct negotiation. (It should be pointed out, however, that most governmental owners would still be placed in this group, even if they were not prevented by law from direct negotiation, because of their size and/or because of a lack of autonomy of and authority in key decision-making personnel.)

Group II Owners

This group includes those owners to whom contractors and other elements in the construction industry are more likely to be responsive because of the distinct possibility of repeat business through direct negotiation. This group would generally include the following:

1. Developers and consistent real estate improvement investors operating within given geographic areas

2. Corporate and private owners (excluding the very large corporation in most instances) who are consistently in the construction market within a given geographic area

There are, of course, some special-situation variations to the foregoing, but generally these groupings will provide the basis for owner evaluation in the predesign analysis.

This book is based primarily on providing design and construction management services for group I owners. However, the design approach certainly, and the other procedures as annotated, are generally applicable to group II owners.

Group II owners are somewhat more variable in characteristics than group I owners. Sometimes a client may appear to be of group II but will be better off when group I owner procedures are followed. It is important, though, to recognize that the above categorization results from only a very broad and superficial analysis, and that every owner has at least some unique characteristics. The following are some further variable characteristics of owners in both groups:

In the developer and investor categories, a number of owners do not maintain a full construction program management capability as part of their company staff. While the majority of developers do probably have good in-house construction management capabilities, many having their roots in the construction field, a more investment-oriented owner in this group may not. For these investment owners, either an architectural/engineering firm (or a construction management company working with the architect/engineer) might well apply productively the full time/cost control system, simply employing direct negotiation for construction contracts rather than competitive bidding. (The time/cost control system forms the major part of the following chapters.) In these instances, the contractor or contractors might be brought in early in the design phase, possibly taking part in the predesign project analysis as described in Chapter 5. This assumes, of course, an evaluation indicating with reasonable certainty that the selected contractor will be truly responsive to the owner, this being basic for the group II classification.

As mentioned above, many developer-owners have good in-house construction management capability. Others rely heavily on favored contractors for construction management services, oftentimes having continuing relationships with such construction companies. In fact, some of the best construction management talent is often found here, in these companies which reap large volumes of repeat business from active and successful developers. In these instances, there is probably little need or place for the separate construction management service company. Of course, architectural and engineering services will be required, but it will probably be the contractor or the developer-owner who has the construction management responsibilities. Under these circumstances, it will still be most productive for the architect, contractor, developer's people, engineers, etc., to hold a predesign project analysis (see Chapter 5). It would seem advisable and most reasonable for the architect to initiate and lead the analysis. Otherwise, under this set of circumstances, the architect will be functioning primarily as the designer. However, many activities similar to those in the time/cost control system will be performed jointly by the architect and contractor or developer's staff. In particular, the relationship between designers and estimators, similar to those described in Chapter 7, will be most productive. And the design approach discussed in Chapter 6 is applicable to any project.

It is also true that private developers and investors as well as other medium- or smaller-size corporate and private sector owners, all being of group II, do not have quite the construction program management problems that most group I owners have in their programs. This is true because of the project size, owner size, and the administrative and regulatory constraints facing most group I owners. While existing with any owner to some degree, those constraints tend to require a much greater management effort for group I owners. One must not overlook, however, that owners in the group II category, particularly developers and investors, face a degree of competitive economic pressure that is rarely as exacting among group I owners except for some larger corporate owners.

OWNERS' STAFFS

The majority of sizable group I and II owners will usually have an in-house facilities management capability which will vary from minimal in some cases, to substantial in many cases, to overstaffing in a number of cases. These groups, which are usually labeled "Facilities Department," "Real Estate and Construction Division," or the like in most United States corporations (Public Building Service, Corps of Engineers, Building Authority, or a public development corporation are examples in government), rarely find it feasible or desirable to maintain groups that carry out detailed project design or construction management, though there are some notable exceptions. Instead, these groups or agencies usually have the responsibility for programming new facilities; developing total funding projections; selecting and negotiating contracts with architects, engineers and construction managers; and generally tracking all of the given company's, institution's, or agency's construction programs. In these instances, the full application of the time/cost control system by the project architect/engineer or construction manager would be in order and should be highly productive, using bidding for construction contracts for group I and usually negotiation for group II.

An important thing to remember, however, in the foregoing set of circumstances, is that the owner's in-house staff must be fully aware of the construction management system being employed, be in sympathy with it, and be kept fully informed. Frequently, higher management will be more aware of the need for and goals of such a system, and the in-house facilities or construction management group may be less willing to change from old methods. This presents a communications problem to the architect/engineer or outside construction manager—one that must be solved for a successful project. It is extremely important for the in-house facilities manager to realize the needs of his own management or agency officials and to recoginze that construction management as well as design are ever-changing fields. Nothing is as valuable as experienced good people, but it is a shame for such people to waste efforts on the wrong approach or to use old methods when new and better methods and systems are available.

Not to leave the wrong impression, however, some of the in-house facilities managers have made most significant contributions to the evolving profession of construction management and construction program management.

As previously discussed, the time/cost control design and construction management system, which is described in the balance of this book, may be brought to bear on a project either through a combined design and construction management service or carried out in conjunction with a separate architect/engineer. Nevertheless, that rare owner who can justify and maintain a complete design and construction management staff in-house may make a full and productive utilization of the system.

There is the strong opinion held by many, of course, that it is uneconomic and unwise for any owner, governmental or private, to attempt to maintain a full architectural, engineering, and construction management group within his own organization. There are certainly strong arguments favoring that view. Among them are difficulties in attracting the best talent seeking lifetime careers in these professions, the absence of competitive pressures, the inexperience of top management in these areas and their orientation to other fields and activities, tendencies toward overstaffing, much higher costs when all overhead and other expenses are taken into account, and uneven work loads. These are but a few of the reasons why most companies and agencies have steered away from this course. Nonetheless, should a given company or governmental agency maintain such a staff, the time/cost control system would be directly and fully applicable for owners in both group I and group II.

Experience has shown that, in any case, a company or agency maintaining in-house facilities management staff would do well to "fish or cut bait." Either the entire design and construction management service should be provided in-house, or it would be best to stick exclusively to those activities referred to earlier in this chapter in the discussion on in-house facilities management capabilities of group I and group II owners, which are more fully listed below:

1. New facilities programming and coordination of facilities requirement projections (including coordination of industrial engineering in the case of industrial owners)
2. Development of project budget and funding projections
3. Selecting and negotiating contracts with architects, engineers, and construction managers
4. Existing facilities' maintenance and services
5. Minor renovations
6. Space management
7. Tracking all construction programs
8. Continuing study of operations as related to efficient and profitable use of facilities

(A number of these activities are, however, frequently enhanced by consultation or services from an outside construction program management firm.)

A great deal of inefficiency and some danger for the owner exists in the great compromise middle ground that lies between the two above approaches of in-house design and construction management versus facilities management. Assuming that the owner will not provide complete design and construction management services, he would then do well to maintain a minimum size, high-caliber staff that would, in turn, control his construction programs through complete and carefully tailored contracts with architects, engineers, and construction managers.

Over the years, the very inefficient situation of the utilization of the "owner's representative" at the jobsite has developed and has frequently caused confusion, unnecessary involvement by the owner in details, and added costs or lost time of very substantial dimensions to the owner. All noncontractor representatives at the jobsite should be those of the architectural and engineering firm or of the construction manager unless the owner is providing all construction management services.

Another practice that has developed in some corporate and governmental owners' organizations has been the concept of maintaining a skeleton design and construction management group and supplementing from outside firms as work load requires. Except for developmental operating studies, this is most unwise as it leads to inefficiency, divided responsibility, and overstaffing.

The owner's in-house facilities staff should not automatically become the arbitrator between the architect/engineer and contractors nor between the construction manager and contractors, architects, engineers, etc. Instead, whichever party has the prime professional service contract with the owner should be considered a part of the owner's organization, and any problems or dissatisfaction with those services should usually be dealt with by the owner dealing directly with his A/E or construction manager and without bringing in any third party. Short of there being dissatisfaction or a problem going unsolved, the owner's in-house facilities staff will operate at maximum efficiency and effectiveness if it will depend on the effectiveness, responsiveness, and responsibilities of the architectural/engineering firm or the construction management company to handle all design and construction management details of the project.

Most owners need to improve the quality and scope of the contracts for professional services, and all need to exercise the greatest of care in the selection of these firms in the first place. And then, the facilities manager should give continuous support, accompanied always with displays of confidence in and reliance on the selected firm, after making clear the goals to be accomplished. Those five things are the facilities staff's most important acts for the individual project.

School of Dentistry, Medical College of Georgia, Augusta, Georgia. (Heery & Heery, Architects and Engineers.)

CHAPTER 4

Introduction to the Time/Cost Control System

All the projects illustrated in the case histories in Chapter 1 were designed and managed using the time/cost control system that is described in the following eight chapters.

An important fact for the reader to understand at the outset is that time and cost are not likely to be controlled as well as may be done when this effort is divorced from the design process itself or when, in fact, time and cost control is not a part of the designer's philosophy.

As was discussed in Chapter 2, in the period of the 1960s and early 1970s, a number of architectural and engineering firms as well as management consultants operating in the construction industry, along with subsidiaries or divisions of several large building contractors, began to develop some more formalized methods and techniques for construction management.

The time/cost control system had its beginnings in 1955 and was fairly well formed as a definable method by 1961. The system was first developed by the author and his associates to meet the needs of corporate industrial owners moving new manufacturing operations into the southeastern United States during the 1950s and 1960s. The system, combined with full, normal, architectural services, was developed as an answer or alternative to the design/build "package." While the design/build approach may offer some distinct advantages, particularly to the industrial owner seeking a very simple new distribution facility or the like, it was often characterized later by some owners who used it as being not unlike going to the doctor and the undertaker at the same time.

The time/cost control system, as well as other similar approaches, may be brought to bear on a project either through a single design and construction management service or through a separate professional construction management contract.

As discussed in Chapter 2, a number of distinct advantages would appear to exist for the owner in using the combined, single, design and construction management contract for the typical building or a group of related buildings in an owner's construction program.

However, use of the construction management firm, which is separate from the project architect/engineer, definitely has many applications that can be highly productive and that can be the best of the owner's options under the right circumstances. These circumstances would include the case in which the selected architect does not have the desire or capability to provide a complete construction management service. Another situation would be the project that consists of a series of related buildings with different architects.

The time/cost control system involves seven basic components that are either additions to or modifications of the typical architectural and engineering service (such as is covered in AIA contract form B-131). They are as follows:

1. Predesign project analysis (Chapter 5)
2. Systems approach to design (Chapter 6)
3. An integrated cost-control system (Chapter 7)
4. Time-control contract provisions (Chapter 8)
5. Scheduling and information system (Chapter 9)
6. Bid and negotiation management (Chapter 11)
7. Management of contracts and construction (Chapters 10 and 12)

An architect/engineer providing a combined design and construction management service would provide all the above along with normal A/E services.

In applying the system through a separate construction management contract, the construction manager would normally provide the services listed under 3, 4, 5, 6, and 7 directly. He would initiate and be a participant in 1. And he would make recommendations to the architect and owner relative to 2.

While it is feasible for the principals of an architectural and engineering firm to train their project managers and other key personnel in this or other similar systems, as it would likewise be feasible for a construction management company, there are some important cautions that should be mentioned:

1. It will take experienced architects, engineers, or construction personnel, in the first place, to be trainable as lead construction managers.

2. No two projects are alike; any construction management system will have to be adapted to the individual project. The time/cost system has always been in a constant state of evolution, as it most assuredly will continue to be.

3. One basic concept of the time/cost control system is that there is not complete reliance on any one system or procedure, such as critical path method scheduling or the utilization of industrialized building systems. Rather, everything that can be reasonably done for the individual project is done. Consequently, it may sometimes appear that one component of the system is not necessary or is redundant. But, to ensure best results, all components of the system should be brought fully to bear on each project except as otherwise recommended in specific cases in Chapters 5 through 12.

4. Another basic concept of the time/cost control system is that the construction contract format should always be kept in its simplest and most definitive form. Therefore, all things being equal, the construction manager should usually be working toward a single-responsibility general construction contract for the owner by the time work starts at the site.

However, there are very definitely certain situations where an alternative approach would be in the owner's best interest or will be dictated by law or circumstances. A further notation on this point: As construction management further evolves and its use becomes more widespread in the United States, the advantages to an owner of the single-responsibility lump-sum general contract will decrease, relatively speaking. And this trend will be accelerated by the further development of industrialized building systems.

The application of the seven components of the time/cost control system, as covered in Chapters 5 through 12, pertains primarily to projects for owners in group I. (See Chapter 3, the section entitled "Evaluating the Basic Characteristics of Owners for Construction Purchasing Posture.") Consequently, some differences for owners in group II will exist. Most of these differences will be evidenced by rereading the abovementioned section in Chapter 3 after completing the balance of the book. They will have to do, primarily, with utilization of direct negotiation instead of competitive bidding, less formalized approaches, and the lack of a role, in some cases, for the separate outside construction manager.

A final general note before going on to the next eight chapters and the details of the time/cost control system: It is probably presumptuous to refer to what follows with the word "system" and its certain mystique or to entitle it in such a way as to make one believe that the time/cost control system contains magic formulas or even pat answers. Instead, the reader should realize that the "system" is simply a series of interrelated, commonsense procedures that have evolved and that have been thoroughly proved to be practical and productive. Probably no one component is completely original. So one should not be surprised to find the components to be fairly simple or, in several cases, to be similar to procedures others have used.

CHAPTER 5

Predesign Project Analysis

The word "predesign" is used here in the sense of being in advance of the normal design process. The basic purpose of the predesign project analysis is to launch the design process on a multidisciplinary and efficient basis, taking into account both design and construction management considerations.

However, the effect of this technique will be far beyond the foregoing in that this basic approach overthrows traditional approaches to the design of a building and therefore will unquestionably have a major impact on the design end product.

Traditionally, design has often been approached through the "fountainhead" theory. This specifically includes the approach employed not only by earlier period architects but also that which has been espoused and employed by many contemporary architects on through the beginning of the 1970s. The theory is that the design of a building, group of related buildings, or other architectural elements must flow from the mind of one man. Preferably, this man is some sort of genius. The designer plays the role of the "fountainhead," and all other design and construction management disciplines have as their roles merely the support and execution of his design concept.

A great deal of very bad architecture has been produced under this theory, though a great deal of poor architecture will probably always be produced with any given approach. But in an emulation of such true architectural geniuses of the twentieth century as Mies van der Rohe, LeCorbusier, Wright, and Gropius, the typical design process in many architectural firms has sometimes deteriorated to something like the following:

The program is given to a designer in the design department, who retires to his ivory tower and with soft pencil and yellow paper in hand starts the one-man process in the full expectation that he alone will come up with the right design. His main areas of attack are floor-planning, siting, and visual effect. He may consider, after some study, that he should seek a limited amount of structural or mechanical engineering advice. And, after a while, he may begin to use one or two draftsmen. On a particularly complex project, he may begin to put one or two more designers to work under his supervision, and he may even direct some research activites by others. After a design begins to take shape, which may well be several weeks or even months into the process, he may then bring in an estimator. Alas, all too often, he then starts over.

This age-old and honored approach is, at best, inefficient. It is surely lacking in any real engineering integration. There is insufficient early construction cost input. Realistic time schedules for the design process are ignored, as are overall constraints that should have early identification. It is assumed, falsely, that construction management considerations can wait until later or be set in motion independently of the design process. Yet the process is a time-honored one, and he who tampers with it is said to be an insensitive poetry smasher by those who would promulgate the fountainhead approach.

But as Caudill says so well in *Architecture by Team,*[3] "The prima donna is dead."

The predesign project analysis is a multidiscipline "think tank." In the typical building project, the participants should include the project manager, project architect, or construction manager (or all three depending on terminology and whether or not there is a separate construction manager); the architectural designer or designers; the structural, mechanical, electrical, and civil engineers; the landscape architect or planner or both; the estimator; and the interior designer. In certain cases, it may even be wise to bring the graphics designer into the picture at the outset. And in some cases, a professional representative of the owner or user can contribute.

In organizing and instigating the analysis sessions, it is important not to let them become committee meetings or mere briefing sessions. If construction management is being handled by the architect under a design and construction management contract, it would normally be the project architect or project manager (again, depending on terminology) who organizes the analysis sessions. If construction management is being handled by a separate group or company, the construction manager may be the organizer.

Once the meetings are organized and set in motion, it will be the principal architect, project architect, project designer, or whoever has the primary architectural design assignment—along with the construction manager (if separate)—who should jointly lead the sessions. Persons leading a predesign project analysis will need experience to develop their own techniques for making the sessions truly productive.

A conducive environment is sometimes hard to create, particularly at first. Each participant, including principal architects and engineers, must be willing to throw out for group consideration ideas that are less than fully developed, and each must display a willingness to accept criticism in constructive ways and with good humor. Participants should be made to understand that, at this stage of the project, comments and creative contributions in areas outside their own respective disciplines are not only acceptable but desirable.

Seemingly logical chronology should be avoided. There should be no order for decision making (i.e., first, design concept; second, engineering systems selection; third, budget analysis; fourth, schedule development; etc.).

Instead, all aspects of the project should be analyzed as they relate to the other aspects. Decisions on major points and concepts should evolve simultaneously. In a few words, these work sessions should be formless and all aspects of the project should be considered together.

Preparation for the sessions should be limited. However, pertinent givens

[3]W. Caudill, et al., *Architecture by Team,* Van Nostrand-Reinhold, New York, 1971.

The predesign project analysis is a multidisciplined "think tank" that simultaneously deals with all aspects of the project and develops basic concepts in both areas prior to the commencement of design work.

should, of course, be put in the hands of participants in advance if available. This material would usually include a program of requirements, site, and surrounding area data. (Close proximity to the site, away from the conventional office environment, is often good for a predesign session.) Basic givens would also include budget information, zoning and agency approval information, and any other obvious and major givens.

However, the separate construction manager who is leading an architectural/engineering group through this process for their first time will need to do somewhat more preparation and planning for the sessions. In these cases, the predesign project analysis sessions may extend over a somewhat longer period of time. There may be an introductory session, a break of several days, and a resumption with additional preparation by all parties.

In addition to accomplishing the other stated goals, the predesign project analysis gives the principal of a larger architectural/engineering firm an excellent opportunity to see that a project is properly launched. It quickly focuses on most major aspects of the project so that the principal may inject his own input before much time and manpower is expended.

A productive predesign project analysis may run from the better part of one day to several days. If the session runs a full work week or longer, it has probably been allowed to run too long to be effective. One or possibly two days is probably typical, usually with some documentation follow-up by the project or construction manager. Only the simplest of projects, which are rare, could be fully covered in a half day.

For a more or less typical project, here is what should be accomplished in a predesign project analysis:

1. *Design concept development* or at least the determination of the direction in which design work should proceed, plus identification of areas of needed research.

2. *Elimination of design blind alleys* to the extent feasible.

3. *Identification of constraints,* including construction, fabrication, and delivery constraints as well as labor, design, site, owner decision and financing, agency approval, and other administrative constraints.

4. *Systems utilization analysis* and decision.

5. *Systems development potentials analysis* and decision in the event available industrialized building systems are not applicable.

6. *Owner evaluation*. See the section in Chapter 3 entitled "Evaluating the Basic Characteristics of Owners for Construction Purchasing Posture."

7. *Construction management plan.*

8. *Critical date schedule* that incorporates the major aspects of the construction management plan. See Figure 5-1.

9. *Budget analysis* or budget development plus identification of areas of major potential cost problems, cost reduction possibilities, and construction purchasing opportunities.

10. *Tentative selection of major engineered systems.*

FIGURE 5-1

CONSTRUCTION MANAGEMENT PLAN

Schedule of Critical Dates/1-8-73
***Project:* Office Building, Atlanta, Georgia**
***Available:* Property Survey, Owner's Program of Requirements**

	1973
1. *AE authorized to proceed.* Commence Predesign Project Analysis.	9:00 a.m. Mon., Jan. 8
2. *Owner authorizes Soil Tests and Topo Survey.* Complete Predesign Analysis.	Tues., Jan. 9
3. Enter negotiations for Elevators, Electric Switchgear, and Roof Insulation (1st Early Bid Packages)	Wed., Jan 10
4. Complete Schematic Design with Total Project Budget including confirmed CCAP[4]; present to Owner along with manufacturers' proposals on 1st Early Bid Packages.	2:00 p.m. Wed., Jan. 24
5. Owner approves schematics, budget and CM Plan and authorizes award of contracts for Elevators, Electric Switchgear and Roof Insulation	Fri., Jan. 26
6. Commence Design Development Phase and preparation of 2nd Early Bid Packages including: Pile Foundations, Structural Frame, Ceiling/Lighting System, HVAC System	Mon., Jan. 29

[4]Construction Contracts Award Price(s)

7. Obtain Jurisdictional Agencies' and Insurance Underwriters' preliminary approval of Schematics	Thurs., Feb. 8
8. Complete Design Development documents and 2nd Early Bid Packages. Present foregoing to Owner.	2:00 p.m. Wed., Feb. 21
9. *Owner approves Design Development and 2nd Early Bid Packages documents and authorizes AE to proceed.*	Fri., Feb. 23
10. Bids received for 2nd Early Bid Packages	4:00 p.m. Tues., Mar. 20
11. *Owner authorizes award of all 2nd Award Contracts.* Permits and insurance commitments obtained on D.D. documents.	Wed., Mar. 21
12. Contracts awarded and NTPs issued for Pile Foundations (and related excavation work), Structural Frame, Ceiling/Lighting System, HVAC System.	Fri., Mar. 30
13. Commence Interior Furnishings Design and Selection	Mon., Apr. 2
14. Complete General Contract Documents and issue for bids.	Mon., Apr. 9
15. Prebid Conference for G.C.	2:00 p.m. Wed., Apr. 18
16. Pile Foundations complete. This contract not to be transferred, but work to be accepted by G.C. bidders.	Thurs., Apr. 19
17. Receive G.C. bids.	4:00 p.m. Thurs., Apr. 26
18. *Owner authorizes 19 below.*	Wed., May 2
19. Award G.C. and Transfer all Early Contracts to G.C. except Pile Foundations.	Fri., May 4
20. Interior Furnishings, Landscape and Graphics designs presented to Owner.	2:00 p.m. Wed., May 9
21. *Owner approves above designs and authorizes AE to proceed.*	Fri., May 11
22. Issue Interior Furnishings, Graphics, and Interior Space Fit-up packages for bids.	Mon., June 25
23. Receive Interior Furnishings, Graphics, and Interior Space Fit-up bids.	4:00 p.m. Thurs., July 19
24. *Owner authorizes award of foregoing.*	Tues., July 24
25. Contracts awarded and NTPs issued for Interior Furnishings, Graphics, and Interior Space Fit-up.	Fri., July 26
26. Landscape package issued for bids.	Mon., Aug. 20
27. Bids received for Landscape work.	4:00 p.m. Thurs., Sept. 6
28. *Owner authorizes award of Landscape contract.*	Mon., Sept. 10
29. Landscape contract awarded and NTP issued.	Wed., Sept. 12
30. Beneficial Occupancy except for Landscaping.	Fri., Nov. 9
31. Furnishings Delivery.	Wed., Nov. 14
32. *Move-in.*	Mon., Nov. 26
33. Final Acceptance with completed Landscaping.	Wed., Dec. 19

As a *minimum,* the following should result from any predesign project analysis:

1. Determination of design direction
2. Identification of major constraints
3. Owner's construction purchasing posture evaluation
4. Tentative construction management plan
5. Critical date schedule
6. Budget givens and limitations

For each project, there is a point of diminishing return in the scope of the predesign project analysis. For example, in the simple, light manufacturing plant, it is often entirely reasonable to evolve a very definitive design concept that would probably go as far as selecting type of structural frame, sidewall construction, bay system size, and building siting. On the other hand, in a multibuilding medical complex, probably only a basic design direction can be arrived at insofar as design concept is concerned.

Constraint identification comes only with experience, but there are some common areas that might be listed here as checkpoints. It is important to note, however, that two areas that are frequently thought to pose constraints rarely do. They are site grading and foundation work. Many owners have been badly misled into open-ended construction contracts, loss of competitive position, and the diversion of critical early-phase design work into "cart before the horse" activities by this particular false assumption on site work. Only in the cases of extensive earthwork, major high-rise structures, or exceptionally complex or extensive excavation or foundation work are foundation work activities likely to constitute a constraint that cannot be dealt with in the normal design and award sequence. Here is a partial list of items that frequently do show up as real constraints:

Structural frame fabrication, delivery, and erection
Site availability and zoning
Utility, street, and site access work
Electric switchgear delivery or other major electric equipment
Labor contract terminations
Major mechanical machinery delivery
Bridge crane delivery
Elevator installation
Laboratory equipment delivery
Furniture delivery
Special and unusual major equipment, its delivery and installation
Agency approvals
Owner's reviews and decisions
Extensive or complicated grading or foundation work (infrequent except as discussed above.)

The construction management plan is simply that, a plan for managing the construction program. As a minimum, it should include the following:

1. Plan as to competitive bidding versus direct negotiation for the construction contract or respective separate contracts.

2. As a result of the constraint identifications, the plan for early bidding or negotiation (if required). This part of the plan, along with the following two items, pertains to what is sometimes referred to as "phased construction."

3. Plan relative to early awards. It should be noted that early bidding does not always mean early award of separate contracts for parts or systems of the building. This situation is found primarily in industrialized systems projects. Here, it is desirable sometimes to identify systems component manufacturers before the completion of general construction documents. It can also be desirable to tie down and hold the price of all of the system's components, which is feasible for longer than normal thirty-day periods with industrialized systems, even though construction contracts for only one component may be awarded early—or perhaps none will be so awarded.

4. Plan relative to transfer of contracts for constraint or systems work to a general contract. Experience with the time/cost control system has indicated that, normally, any contract that will still be incomplete by the time of award of the general contract should be transferred into the general contract.

By the time the predesign project analysis has been brought to a conclusion, designers representing all the major disicplines should be able to move forward earlier and more productively. The multidisciplinary team should have become a visible and cooperating force. Mutual respect should have been established. And most important, those forces that are likely to produce architecture that is sympathetic to the medium and relevant to the user's needs will have been set in motion.

CHAPTER 6

Systems Approach to Design

Design is the basic business of the architect as well as the engineer who deals with environmental and structural systems. In large measure, design is also the concern of the construction manager, for he must be able to give early guidance to the designer, when the construction manager is operating as a separate professional adviser, in the areas of cost, methods of construction, sequencing of construction, and other matters. Consideration of these factors is inherently a part of the design process.

All other activities of the architect/construction manager must be successfully factored into the overall approach and set of services, including design, that form the process that creates present-day architecture.

But design is basic to all, the common denominator in any effort to create buildings, urban complexes, and other facilities. Design, after all, determines the degree of almost all the long-term benefits of the building program as well as having the major direct effect on cost and, in no small measure, on time. Design is the major determinant of original and operating cost. Design is the basic tool available for cost control.

The term "systems approach to design" can be properly applied and understood to have several meanings.

It means combining sophisticated methods of decision making (often computer-assisted) in the planning and design process with a philosophy of setting design and development in motion based on criterion and performance requirements. (The latter should not be taken to mean that "performance" specifications are always a part of a systems design effort or necessarily part of the utilization of existing industralized building systems.) This combining of methods and philosophy is fundamental to the systems approach and is surely compatible with the author's design philosophy and the suggestion herein of a redirection in design thinking. It is also compatible with the idea that the designs that are most relevant to owner/user needs are likely to come out of a highly integrated design and construction management system.

Insofar as systems hardware is concerned, it is important to review those fundamental reasons why some architects, owners, and other parties (such as the Educational Facilities Laboratory of the Ford Foundation) have been interested in industrialized building systems. Those reasons are *cost* and *time control,* both within desired levels of quality and functional capabilities.

In the 1960s, as the systems movement became more popular like so many other movements, the phenomenon of adopting "hot" words such as "systems" or "systems approach" as jargon occurred, and there ensued the usual myriad misunderstandings, misapplications of terminology, and disorientation of a number of efforts, all under the label of "systems." Some very fuzzy thinking often occurred, with many losing sight of the basic goals of improved

Martin Luther King, Jr., Middle School, Atlanta, Georgia.
(Heery & Heery, Architects & Engineers.)

Administration Building, Woodruff Medical Center, Emory University, Atlanta (model). (Heery & Heery, Architects & Engineers.)

The primary role of this building is to provide a focal point for a distinguished medical center. Cost overrun problems were encountered when the architect's emphasis on traditional design considerations caused a diversion from the integrated cost control system during an unsettled and highly inflationary period in the construction industry. The integrated cost control system has two components that should continue in parallel throughout the design phases (see Chapter 7): (1) Conventional unit price estimating in the early phase replaced by quantity survey take-off estimating in the later phases, and (2) Component budgeting, utilizing composite units and continuous re-checks. In this project, early award work was within estimates, but the general construction package bids ran substantially over budget. This kind of occurrence can not only lose some of the time gained by phasing (see Chapter 10) but can be very frightening to an owner sitting with part of the building bought. In this project, the general construction package was reduced through negotiation to a figure acceptable to the owner, and in the final analysis, phasing still reduced inflationary effect on early award work and saved some time in the overall schedule. It saved a substantial amount of time as compared with a situation of comparable bid overage on a single-contract general contract package.

functional capabilities and time/cost control through a systematic planning and design process, industrialization, coordination, and elimination of redundant engineering.

Figure 6-1—reprinted with permission of *Architectural Record*—illustrates some of the forces at work during the 1960s, forces that have continued since, that would logically direct any given building operation or component's fabrication to be accomplished by industrial methods. The use of such industrial methods assumes, of course, a cost analysis indicating that the industrialized approach will provide either lower cost or approximately the same cost as conventional methods.

Any successful design must always be the result of creative minds solving the given problems of function and environment. Additionally, the systems approach to design has as one basic goal the industrialization of building construction.

Recognizing architecture as an art, not just a visual art but also the art of planning, of translating needs of society to physical facilities, and of management, successful architecture, like any other art, must first and foremost represent a creative and sympathetic treatment of the medium. And in architecture, design must be relevant to the needs of the owner and society, which, in fact, are parts of the medium.

It follows then that the medium must be identified and that the architect and his fellow team members must constantly participate in its evolution. In

FIGURE 6-1

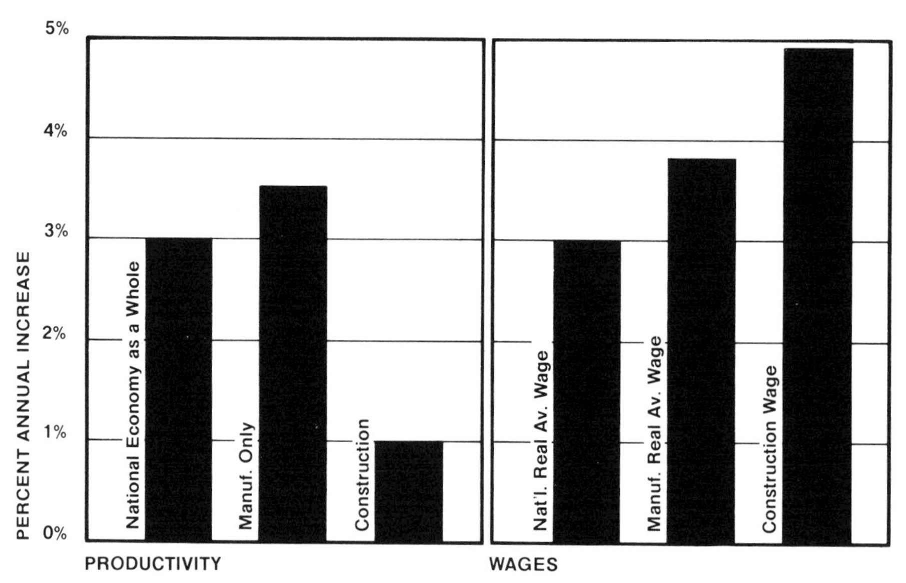

PERCENT OF ANNUAL INCREASE PRODUCTIVITY VERSUS WAGES. 1960-1969

Information Source: ARCHITECTURAL RECORD FEBRUARY '70

the last half of the twentieth century, the major part of the medium is the construction industry.

In the time/cost control design and construction management system, a series of priorities is established as part of the design process as a logical extension of the above philosophy. The priorities are as follows:

1. *Utilization of an available system.* Does the project lend itself to the use of a readily available, multicomponent industrialized building system? If so, that would be the first priority in making selections for construction methods and materials for the project.

2. *Systems development.* If the answer to 1 is no, does the project have characteristics indicating the desirability of a systems development effort? If so, this would be the next priority.

3. *Existing subsystem.* If the answers to 1 and 2 are both no, does the project lend itself to the use of one or more available industrialized subsystems? To the extent that this is the case, then the subsystems would be utilized.

4. *Maximize offsite fabrication and use of standard products.* In all cases, to the extent that it is consistent with owner/user needs, economy, project schedule, and environmental and design goals, maximize the use of offsite fabrication and standard building products.

5. *Orderliness in the design.* To the extent that it is compatible with the project goals, work for orderliness in the design in all of its parts.

The foregoing five points and the preceding discussion are the major aspects of the systems approach to design. The balance of this chapter is given over to discussing the "hardware" aspect of the systems approach, namely the use in design and construction of industrialized building systems.

Since before the design and prefabrication of the Crystal Palace in London in 1851, architects and others have frequently manifested an underlying philosophy that there should be a system in architecture and that the best architecture that can be produced at a given time will make optimum use of the state of industrialization at that time.

After World War II, in England and the United States, and in one or two other countries, there began to be developed a number of systems of construction that made extensive use of preengineering and prefabrication, modular coordination, and systematic compatibility between some of the basic subsystems (structural, mechanical, electrical, architectural). This took place primarily for certain high-volume building types such as schools and housing. From these efforts came both "hardware" and "software," hardware being the physical components, and software being the approach and techniques for developing, designing, utilizing, and managing projects for "systems buildings."

A suburban office park. (Heery & Heery, Architects & Engineers.)

As systems "hardware" has continued to evolve, development has taken place along several different basic directions, as follows:

The Erector Set. The main erector set that has been available in the United States, starting in about 1965, has been the School Construction Systems Development (SCSD) system[5], a multicomponent building system developed first for Southern California schools through Stanford University under a grant from the Educational Facilities Laboratory. There are several versions of SCSD now, and there is a good bit of confusion about their evolution and reliability. The original system was the product of a competition among manufacturers. As time has passed, some of the less flexible aspects of the original system have been modified in favor of more competitive sources for the components, and a wider range of materials and designs has been incorporated within the basic requirements of the system.

The system has been found useful in all parts of the United States, not only for schools but also for other one- and two-story air conditioned buildings such as office buildings, research centers, and certain types of industrial facilities.

[5]John R. Boice, *A History and Evaluation of SCSD,* Building Systems Information Clearing House, Educational Facilities Laboratory, Menlo Park, Calif.

WALT DISNEY WORLD

Administrative Services Office Building for Walt Disney World, Orlando, Florida. (Heery & Heery, Architects, Engineers and Construction Managers.)

This project illustrates a typical "erector set" (SCSD) project of the early 1970s, wherein an existing industrialized building system was combined with modern construction management techniques. The resulting project, which was designed and built in approximately one year, is a high-quality office building providing flexible and pleasant work space at a very low cost. Flexibility in subsystems included ceiling/lighting, HVAC supply and return, and interior partitioning.

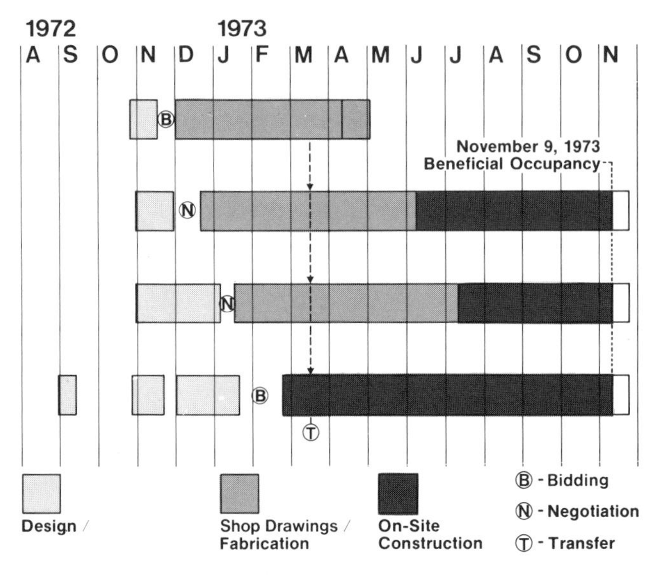

COST DATA

Subsystem components	
Structural frame, second floor deck & roof deck	$157,491
Heating, ventilating & air conditioning	178,384
Ceiling/lighting	126,284
Subtotal of subsystem components	$462,159
Plumbing & misc. mechanical	99,426
Electrical	249,580
Sprinkler	42,200
Building shell (exterior walls, interior partitions)	293,450
Elevator	9,925
Miscellaneous (finishes, sitework, folding partitions, slab, specialties, etc.)	499,540
Subtotal	$1,656,280
Remaining change order (estimated)	26,927
Total	$1,683,207
Size	81,586 sq. ft.
Cost per square foot (incl. estimated change order)	$20.63
Bids received	February, 1973

While some voices have confused the picture by stating that SCSD no longer exists as a true system, the fact is that time and cost may be saved by using the derivative subsystems, whose status in 1974 is as follows:

1. *The structural frames* may be competitively purchased, as or compatible with a 5- by 5-foot plan grid for one- and two-story configurations with bay sizes from 20 by 30 feet up to 40 by 60 feet and 20 by 115 feet. They may be purchased and specified with greatly minimized engineering costs and time by utilizing performance, prescriptive, or proprietary specifications. They are available throughout the United States in steel and in parts of the country in precast concrete. The cost of the frame taken alone is approximately the same as that of conventional construction, with a downward trend relative to conventional construction. Design time is shorter; erection time is shorter; fabrication time is shorter; design cost is lower; designs are frequently pretested and can be accompanied by additional engineering responsibility of the manufacturer; and the system is compatible with the other listed systems assuming that compatibility is required in the individual project specifications.

2. *The HVAC (heating, ventilating, and air conditioning) system* is now available from a number of leading manufacturers of heating and air conditioning equipment and may be purchased and specified on a performance, prescriptive, or proprietary specification. While the original HVAC system was limited to self-contained roof-mounted units, the HVAC system now may consist of either self-contained roof units or roof air-handling units with central heating and chilling units, giving the advantages in certain areas of the lower operating costs of a central station system. Flexible, circular, fiberglass ductwork and lay-in supply and return air terminals along the 5-foot grid line provide flexibility and compatibility with the other subsystems. Design time is shorter; design cost is less; owner's risk is reduced by the added responsibility in the area of engineering accepted by the manufacturers; fabrication and erection time is shorter, and mechanical/structural/electrical conflicts during the design and construction are almost completely eliminated.

3. A large number of manufacturers produce *ceiling/lighting systems* both on the 5-foot grid and on 30-inch multiples, all compatible with the HVAC system, as well as the structural and partitioning systems. Erection time is shorter; quality control is higher; and effects of lighting and reflection are better documented through testing programs of total systems in place.

4. A large number of manufacturers now competitively produce a number of different types of *movable partitions* and demountable partitions that are compatible with all the above systems. A wide range in quality and materials is available. Erection is faster, quality control is higher, and there is a singularity of responsibility for the partition components and their installation.

In addition to the four major subsystems listed above, other subsystems may even be purchased that are compatible with the SCSD system, such as school cabinetry in the case of the programs that have been promulgated in the state of Florida by the Florida State Department of Education.

Figure 6-2 illustrates the basic components of the SCSD system that were first available from the competition winners, and Figure 6-3 illustrates an alternative system with components resulting from interindustry competition on performance specifications.

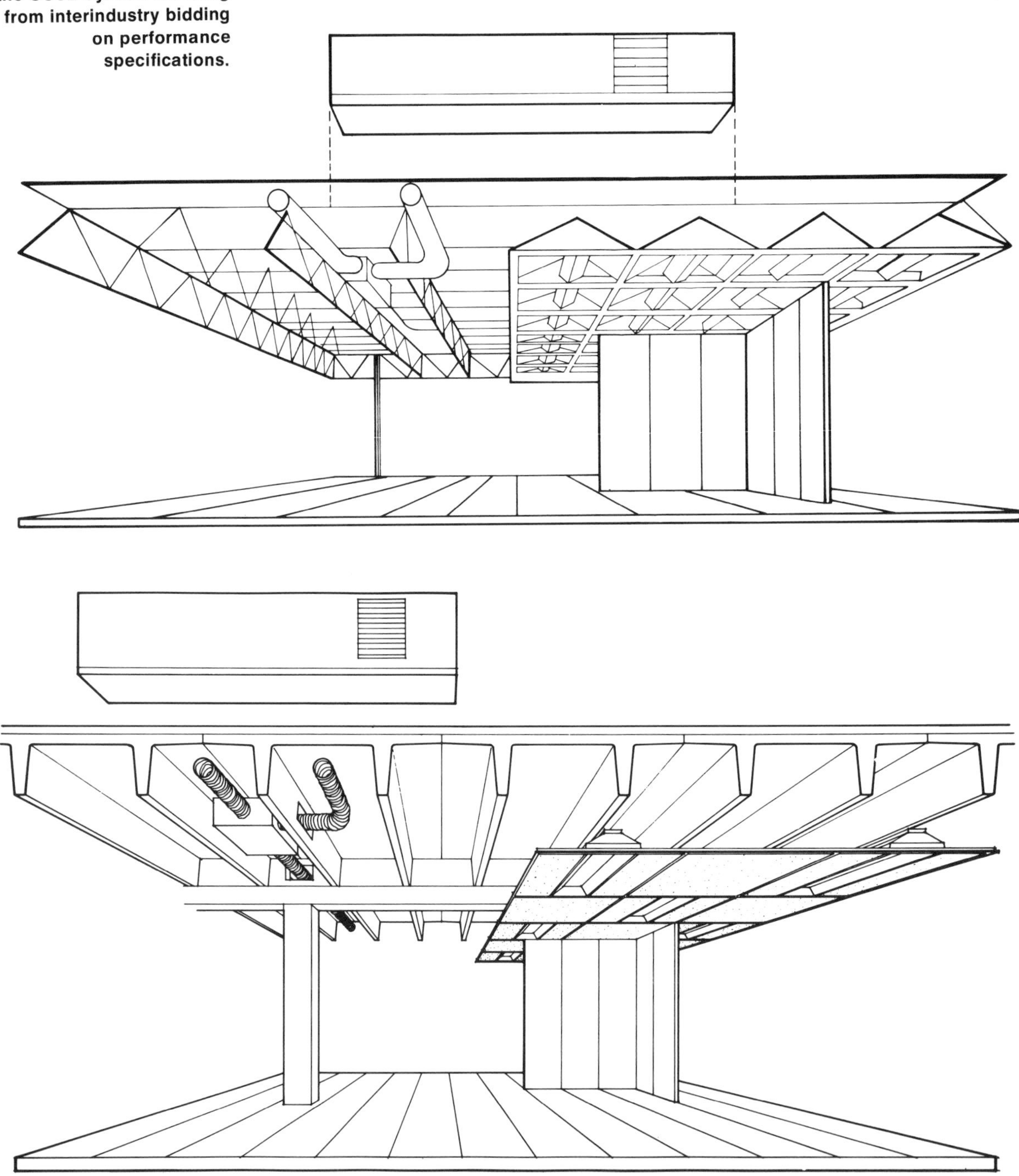

FIGURE 6-2 (Below) The SCSD industrialized building system. Reprinted with permission of the Educational Facilities Laboratory.

FIGURE 6-3 (Bottom) A precast concrete version of the SCSD system resulting from interindustry bidding on performance specifications.

Volumetric Modular Units Growing primarily out of the modular home and mobile home industries, the growth rate of industrialized volumetric modular systems, primarily for housing, accelerated greatly toward the end of the 1960s and in the early 1970s. Figure 6-4 illustrates several different configurations envisioned in a feasibility study for air-force-base officers' and enlisted men's housing. For these projects, performance specifications on which a number of different volumetric modular unit manufacturers could bid were utilized.

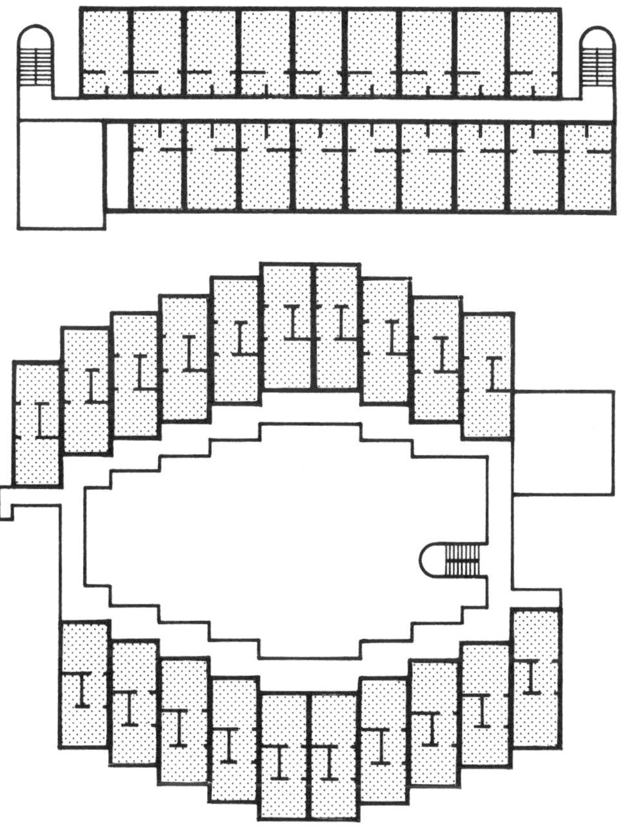

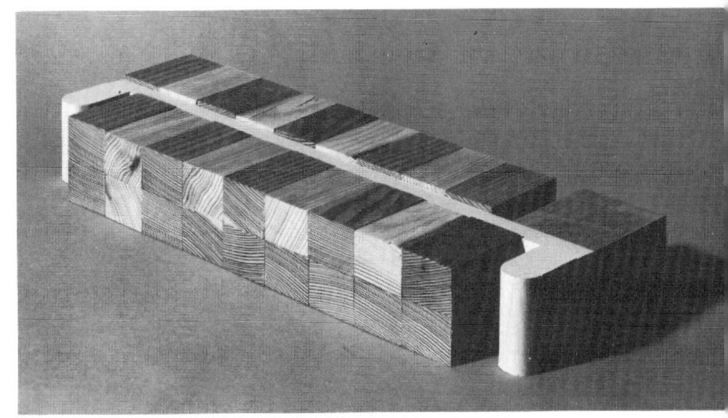

FIGURE 6-4 Several different configurations of volumetric, industrialized modular units envisioned in a feasibility study for air-force-base officers' and enlisted men's housing. (Study by Heery & Heery, Architects & Engineers.)

Figures 6-5 and 6-6 illustrate two of the several resulting buildings after competitive proposals based on performance specifications were obtained.

A wide range of volumetric modular units, varying in construction type, size, plan, function, etc., are available throughout the United States on performance, prescriptive, and proprietary specifications. This segment of the industry, in 1974, is not yet as mature or stable as the segment producing "erector sets," but growth in both areas is a clear pattern.

Panelization Most panelized systems that began to be more widely available in the late 1960s and early 1970s dealt primarily with high-rise housing, using mostly precast concrete and limiting the system primarily to structure and some interior subdivisions. Most have well worked out provisions for mechanical and electrical systems. A number of these systems are available only on a prescriptive (patented) basis, but many can also meet broader performance and prescriptive specifications.

Figures 6-7 and 6-8 illustrate the Thomas Concrete Products and the Techcrete systems.

FIGURE 6-5 (Left) One example of actual results obtained when technical proposals were received, based on performance specifications, pursuant to the feasibility study mentioned in Figure 6-4. Proposer: Community Science Technology Development Corp.; Manufacturer: National Modular Systems, Inc.; Designer: David Soloman, AIA; Program Managers: U.S. Corps of Engineers, Ft. Worth Office, and Heery & Heery, Architects & Engineers.)

FIGURE 6-6 (Left) Another example of results obtained in the Air Force Industrialized housing program mentioned in Figure 6-5. (Proposer: Algernon Blair, Inc.; Manufacturer: Otis International; Designer: Keeva J. Kekst, Assoc., AIA; Program Managers: U.S. Corps of Engineers, Ft. Worth Office, and Heery & Heery, Architects & Engineers.)

FIGURE 6-7 Precast Systems, Inc Thomas Concrete Products, Oklah City, Oklahoma. (Benham-Blair & Affiliates, Inc., Architects and Engineers.)

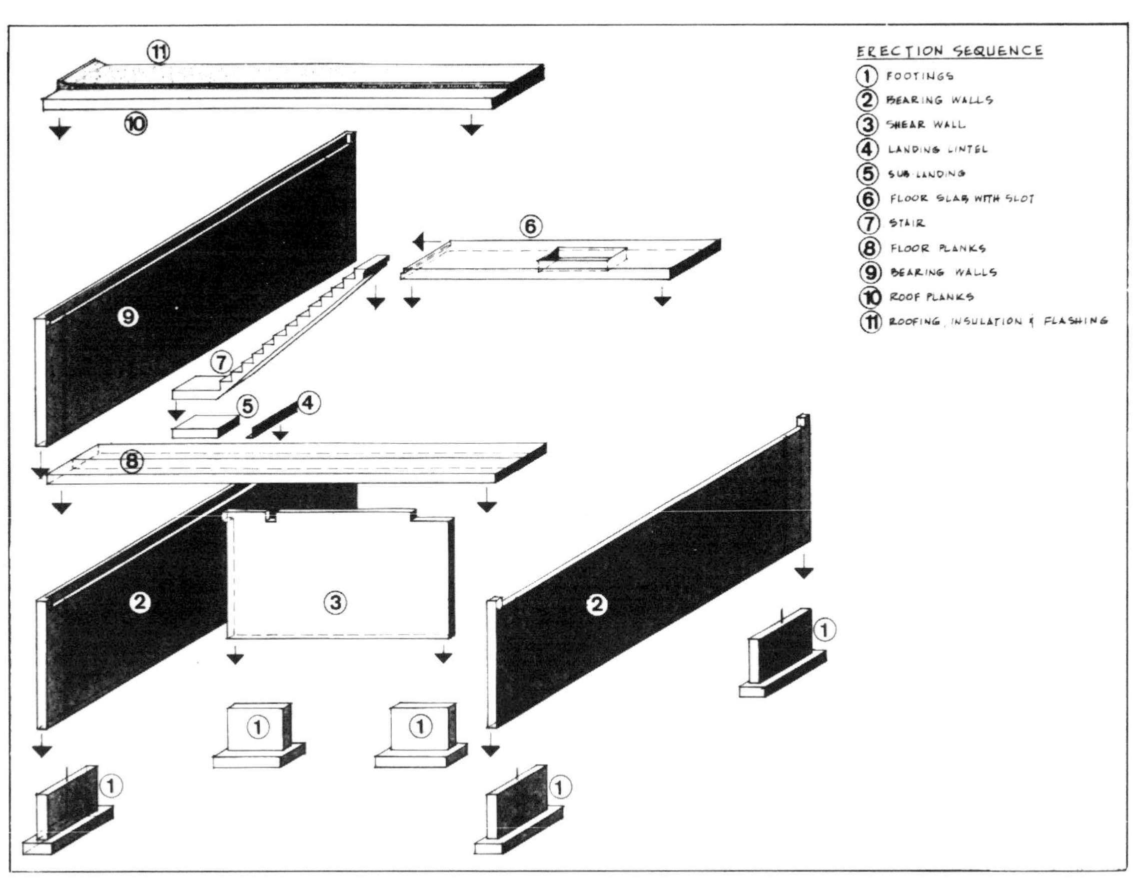
ERECTION SEQUENCE
1 FOOTINGS
2 BEARING WALLS
3 SHEAR WALL
4 LANDING LINTEL
5 SUB LANDING
6 FLOOR SLAB WITH SLOT
7 STAIR
8 FLOOR PLANKS
9 BEARING WALLS
10 ROOF PLANKS
11 ROOFING, INSULATION & FLASHING

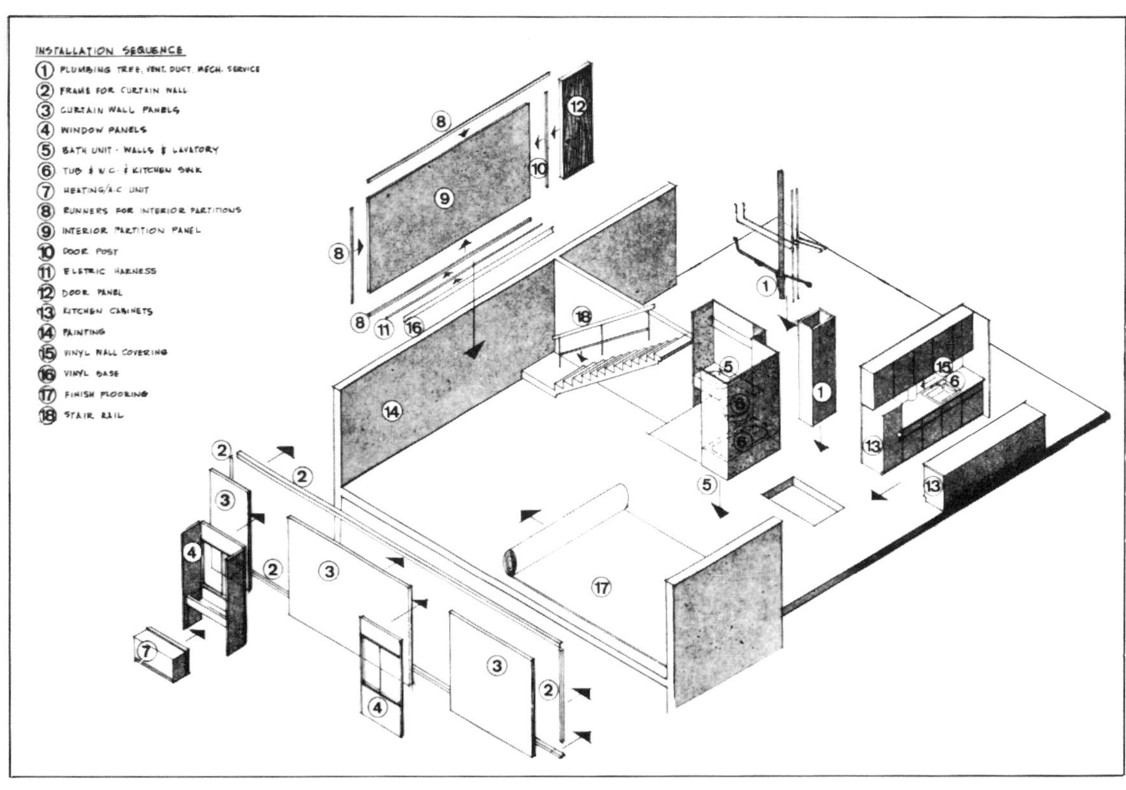
INSTALLATION SEQUENCE
1 PLUMBING TREE, VENT. DUCT. MECH. SERVICE
2 FRAME FOR CURTAIN WALL
3 CURTAIN WALL PANELS
4 WINDOW PANELS
5 BATH UNIT - WALLS & LAVATORY
6 TUB & W.C. & KITCHEN SINK
7 HEATING/A.C UNIT
8 RUNNERS FOR INTERIOR PARTITIONS
9 INTERIOR PARTITION PANEL
10 DOOR POST
11 ELETRIC HARNESS
12 DOOR PANEL
13 KITCHEN CABINETS
14 PAINTING
15 VINYL WALL COVERING
16 VINYL BASE
17 FINISH FLOORING
18 STAIR RAIL

FIGURE 6-8 (Left) Techcrete Systems. (Carl Koch and Associates, Architect. Sepp Firnkas, Engineer.)

Individual Project Systems Development Programs Among the fourteen case histories in Chapter 1 is the group of projects for the New York City Parks Department. Illustrated with the case history information is the precast concrete modular system for the structure of those projects, typical of a simple systems development program on a single project of sufficient volume requirements. In those projects, with the constraint identified as the site availability—causing the critical path to move through the early off-site fabrication and on-site erection of the building structural frame and enclosure—and with other architectural and functional considerations for these fifteen projects located throughout the five boroughs of New York City, the modular systems approach was clearly indicated in the predesign project analysis. A cabanalike, rugged, precast concrete module was the solution concept. The design illustrated in the first case history was the result.

Normal fees for systems projects, as compared with conventional construction, would still seem to be in order, lower engineering cost being offset by the additional administrative and management services that are required of a construction manager or architect/engineer for a systems project. Time saving, along with cost/quality advantages, accrue to the owner, since these additional services may go on concurrently with the somewhat shortened design sequence. In any case, professional services for systems projects are at least as valuable as the services in connection with the conventional construction, and therefore a construction manager or owner should not expect significantly lower or higher fees from the architect/engineer for a systems project. The illustrations discussed below, however, are specimens that will not only provide some guidance but illustrate the brevity and simplicity of preparing contract documents for some systems components.

Figure 6-9 is a complete bid document drawing required for a SCSD structural frame proposal. Figures 6-10, 6-11 and 6-12 are three different versions of specifications that might be used in connection with the structural frame bid document drawing, namely for performance, prescriptive, and proprietary specifications.

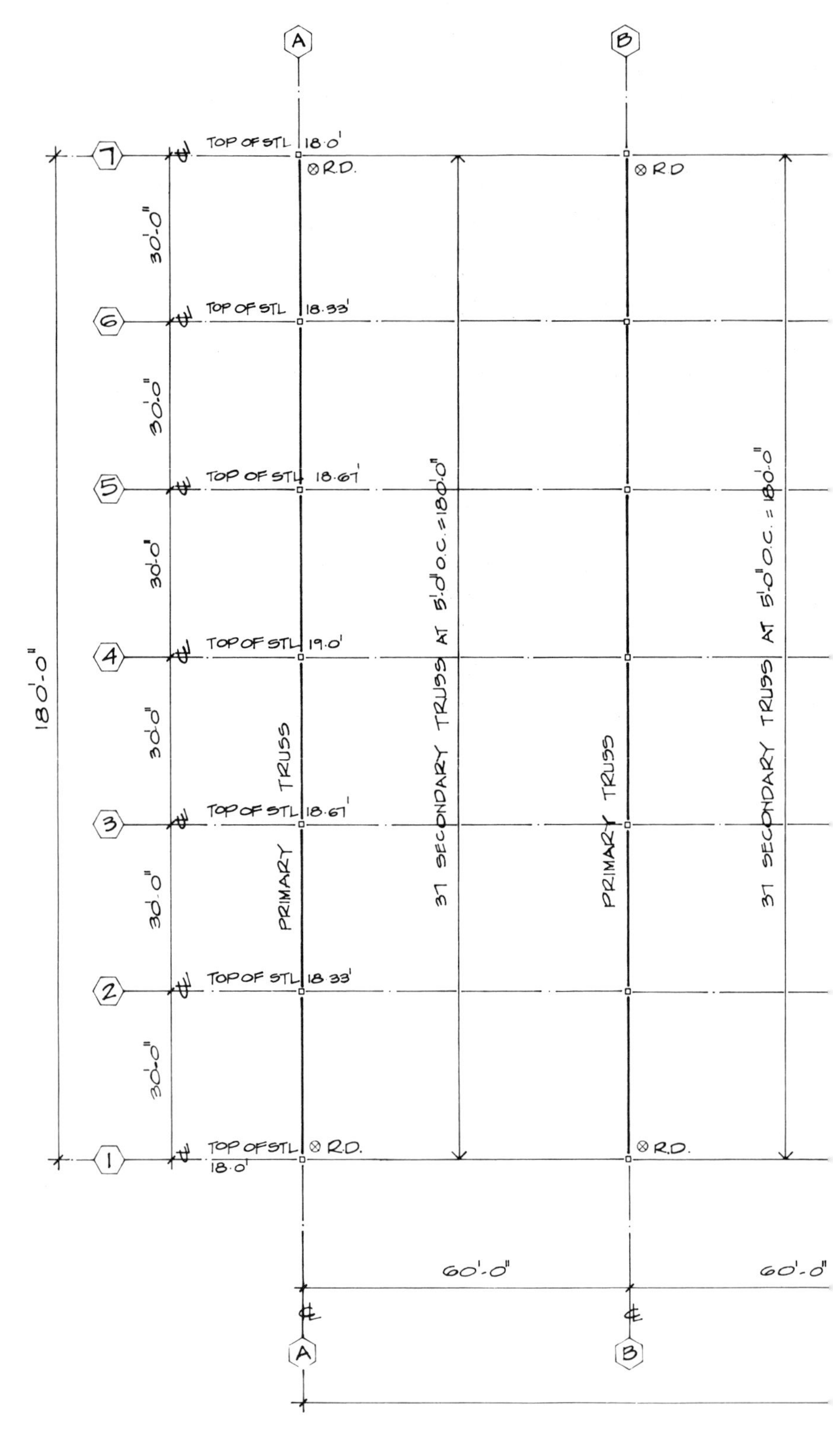

ROOF FRAMING PLAN

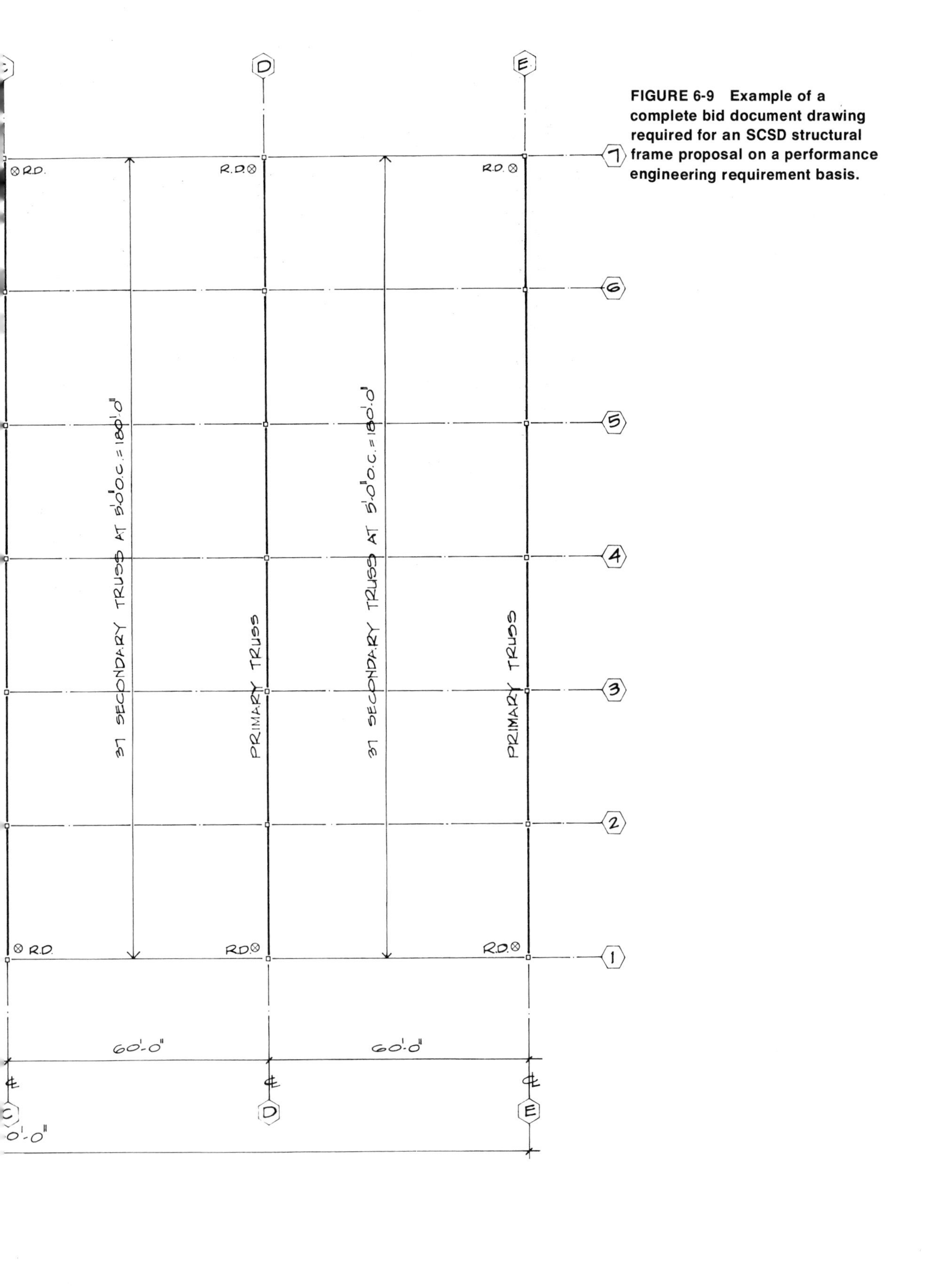

FIGURE 6-9 Example of a complete bid document drawing required for an SCSD structural frame proposal on a performance engineering requirement basis.

FIGURE 6-10

STRUCTURAL STEEL COMPONENT SYSTEM (PERFORMANCE)

DESCRIPTION

INCLUDED:

All items and components forming any portion of the Structural Steel Component System including, but not limited to, the following:

1. Floor and roof decking
2. Secondary and primary spanning members, including bridging if applicable
3. Columns, including base plates
4. Supply of anchor bolts with layout information
5. Wind bracing
6. Special conditions such as expansion joints and openings in floors and roofs
7. Prime coat painting and field touch-up

NOT INCLUDED:

1. Items or components related to Structural Steel Component System which are included as a part of other building systems or components
2. Installation of anchor bolts, which shall be done by others
3. Grouting of base plates, which shall be done by others

PERFORMANCE REQUIREMENTS

ENGINEERING:

Submit a complete set of component drawings, calculations, engineering details, column loadings, and specifications, prepared by a professional Structural Engineer, licensed by the State of _____, and bearing the Structural Engineer's stamp. The Contractor shall be responsible for code compliance and complete structural engineering of the system.

DEAD LOAD:

1. Weight of components self-weight
2. Insulation and roofing: 7.5 psf
3. Ceiling: 3.5 psf
4. Normal mechanical ducts, sprinkler lines and services (verify with compatible HVAC subsystem): 10 psf

LIVE LOAD:

1. Roofs: As required by _________ Building Code.
2. Concentrated Load: 250 pounds at any single truss panel point.
3. Mechanical Load: Coordinate with HVAC SubSystem bidders.
4. Uplift: As required by _________ Building Code.
5. Special Loads: As noted on drawings. The structural system shall be capable of resisting the superimposed wind load through column or frame bending resistance; i.e., without the addition of any shear walls or wall bracing.

JOINTING:

1. Loading conditions to be considered in the design of joints and connections are service loads, including wind, volume changes due to shrinkage, creep and temperature change, erection loads, transportation and storage of members.

2. Consideration shall be given to expansion and contraction, and all solutions shall be related to the 5′ x 5′ planning module. Expansion joints shall occur on column lines without the use of double columns unless specifically shown on drawings.

TOLERANCES:

The structure shall be designed and fabricated to conform in the completed state to the following dimensional tolerances, unless more stringent requirements are imposed on the Structural Steel Component System by other systems.

1. Variations in vertical or horizontal lines and surfaces:
   (a) Columns: in 10′-0″, ¼″; overall, ¼″.
   (b) Primary and Secondary Spanning members: in 5′-0″, ¼″.

2. Variations from straight or from correct position in place, for columns and spanning members:
   (a) In any bay or 20′-0″ max, ½″.
   (b) In any 40′-0″ or more, ¾″.

3. The drift of the frame due to horizontal forces shall be limited to 0.002 times the floor to roof height.

CAMBER:

Secondary spanning members shall be cambered so that they do not deflect below the horizontal under full live and dead load.

INTERFACE:

Compatibility: Design structural system to acknowledge and be compatible with other systems and subsystems. Coordinate all structural components within the system and with adjacent or pertinent building components to assure workable details, connections, clearances, and tolerances.

Roof Drains: Provide openings, deck reinforcing, and sump pans where indicated on drawings to allow for installation of roof drains provided by others.

Skylights and Smoke Vents: Provide openings and deck reinforcing where indicated on drawings to allow for installation of 5′-0″ square skylights and smoke vents provided by others.

ERECTION:

System contractor shall be responsible for the erection and the safe stability of the frame until the entire system is completed and capable of resisting all design forces.

Make adequate provisions for erection stresses and for sufficient temporary bracing to keep the system plumb and in true alignment until completion, including the roof deck and the elements which are a part of the wind resisting system.

Erection operations, including the installation of temporary shoring, shall be carried out without loading portions of the system frame and nonsystem construction in excess of their safe load-carrying capacity. Where the load capacity of the structure may be exceeded, shore constructed portions of the structure, as required, to positive foundation support.

Setting Base Plates: Set column bases on reinforced concrete foundation or piers true and level at the proper elevations. Use methods of vertical and horizontal adjustments during erection which permit full grouting.

Structural members may be stockpiled provided they have been previously inspected and accepted by such inspection agency as may be appointed by the owner and provided storage is done in a manner which will not cause any damage to or will otherwise impair the strength or the finish of the members.

After erection, all welds, bolts, and abrasions in shop paint shall be painted with shop primer.

FIGURE 6-11

STRUCTURAL STEEL COMPONENT SYSTEM (PRESCRIPTIVE)

WORK DESCRIPTION:

1. This section of the specifications and related drawings describe requirements pertaining to the Structural Steel Component System.

APPROVED MANUFACTURERS:

2. The following manufacturers are approved for the structural steel component system:
   (a) Butler Space Grid System
   (b) MaCombber Modular Component System
3. Equivalent products of other manufacturers may be proposed for consideration under the conditions as set forth in the Supplementary Conditions section of these Specifications.

WORK INCLUDED:

4. Columns including anchor bolts, leveling plates, base plates, truss connectors, and all truss connection bolts as required. Anchor bolts and leveling plates shall be supplied under this section but installed as part of work of another section of these specifications.
5. Primary trusses, including all bolts and connectors as required.
6. Secondary or framing trusses, including all bolts and connectors as required.
7. Distribution bracing which shall provide lateral stability to the truss system.
8. Roof deck/diaphragm: Roof deck shall be 1 ½ inches deep, prime-painted or galvanized steel roof deck similar to Inland Type-B, Robertson Section 3, or equivalent, welded to the top chords of the trusses so as to provide a rigid diaphragm. Deck shall be capable of supporting superimposed dead and live loads as noted below but in no case shall deck be less than 22 gauge.
9. Painting (prime coat) of all components and exposed metal surfaces including field touch-up of paint mars and welds.

WORK NOT INCLUDED:

10. Miscellaneous metal work as noted in another section of these specifications.
11. Installation of column anchor bolts.

12. Sheet metal or structural steel not specifically a part of the Structural Steel Component System.

SHOP DRAWINGS:

13. Submit a complete set of component drawings, calculations, engineering details, column loadings, and specifications, prepared by a professional Structural Engineer, licensed by the State of ________, and bearing the Structural Engineer's stamp. The Contractor shall be responsible for code compliance and complete structural engineering of the system.

MATERIALS:

14. Structural steel shall conform to ASTM Designations which particularly describe requirements for each type and grade of steel that is required. Steel that is not particularly designated for type shall conform to ASTM Designation A36. All steel shall be of American manufacture, new and free from defects impairing strength, durability, appearance and function.

15. Bolts, rivets, welding materials, and other fastening materials shall conform to applicable provisions of current edition of AISC Specifications, and American Welding Society.

SHOP AND FIELD PAINTING:

16. Clean surfaces of loose scale, rust, oil, dirt and other foreign matter, immediately prior to painting, in accordance with SSPC - SP2 or 3.

17. Give all surfaces except those to be galvanized, field welded or embedded in concrete, a full-bodied shop coat of paint, in accordance with SSPC - Paint 2-55T or 3-55T. Give surfaces that are to be field-welded a protective coating of linseed oil at shop. Surfaces that will be inaccessible after erection shall have one additional coat of paint of different color than first coat.

18. After erection is complete, touch up parts where paint has rubbed off and paint exposed parts of welds, bolts, washers, and nuts, using same type paint used for shop coat.

19. Leave entire work in proper condition to receive final field painting, which will be executed as part of work of another section.

DIMENSIONAL REQUIREMENTS:

20. All dimensions shall be standardized and all component elements shall be positioned so that their center lines conform to a 5′-0″ planning module.

(a) Primary trusses 60′-0″
(b) Secondary trusses 30′-0″
(c) Columns: Columns shall be detailed so as to provide adequate clearance for compatible ceiling/lighting systems and so as to provide a distance of 12′-0″ clear between finish concrete slab and finish ceiling surface. Clearance between finish concrete slab and top of column base plate shall be a minimum of 4 inches.

PERFORMANCE REQUIREMENTS:

21. The component structure shall be capable of meeting the following requirements:

(a) Superimposed Live Load 20 psf
(b) Dead loads:
Roofing and roof insulation 7.5 psf

Ceiling/lighting	3.5 psf
Sprinkler system and mechanical service lines	10 psf
Fascia line load	40 plf
Self-weight of structure	As required

(c) Wind Load: 23 psf

(d) Uplift 30 psf

(e) The structural system shall be capable of resisting the superimposed wind load through column bending resistance; i.e., without the addition of any shear walls or cross bracing.

(f) Mechanical Equipment Load: As part of the work of this section, provide beams to support roof-mounted mechanical equipment.

(g) Design shall meet all requirements of the ________ Building Code.

FABRICATION:

22. Execute fabrication in accordance with drawing requirements and in accordance with applicable provisions of current edition of AISC Specifications.

SHOP AND FIELD WELDING:

23. All welding shall fully comply with requirements of American Welding Society's "Code for Arc and Gas Welding in Building Construction," as amended to date. Welding operators shall show evidence of satisfactory qualification within twelve months prior to work on this project.

BOLTING:

24. Generally all bolting shall conform to applicable provisions of AISC Specification for the Design, Fabrication and Erection of Structural Steel Buildings, sixth edition. Where structural joints are made using high-strength bolts with hardened washers and nuts tightened to a high tension, the materials, methods of installation and tension control, type of wrenches to be used, and inspection methods shall conform to Specifications for "Structural Joints Using ASTM A325 Bolts," as approved by the Research Council on Riveted and Bolted Structural Joints of the Engineering Foundation.

ERECTION:

25. System contractor shall be responsible for the erection and the safe stability of the frame until the entire system is completed and capable of resisting all design forces.

26. Make adequate provisions for erection stresses and for sufficient temporary bracing to keep the system plumb and in true alignment until completion, including the roof deck and the elements, which are part of the wind resisting system.

27. Erection operations, including the installation of temporary shoring, shall be carried out without loading portions of the system frame and nonsystem construction in excess of their safe load carrying capacity. Where the load capacity of the structure may be exceeded, shore constructed portions of the structure as required to positive foundation support.

28. Setting Base Plates: Set column bases on reinforced concrete foundation or piers true and level at the proper elevations. Use methods of vertical and horizontal adjustments during erection which permit full grouting.

29. Structural members may be stockpiled provided they have been previously inspected and accepted by such inspection agency as may be appointed by the Owner and provided storage is done in a manner which will not cause any damage to or will otherwise impair the strength or the finish of the members.

FIGURE 6-12

STRUCTURAL STEEL COMPONENT SYSTEM (PROPRIETARY)

WORK DESCRIPTION:

1. This section of the specifications and related drawings describe requirements pertaining to the Structural Steel Component System.

MANUFACTURER:

2. Structural steel component system shall be Butler Space Grid System as manufactured by Butler Manufacturing Company, Kansas City, Missouri.

SCOPE:

3. This system contractor shall provide all labor, materials, and equipment to complete the Structural Steel Component System described herein.

4. Work included:

Butler Space Grid Structural System

(a) Fixed base columns
(b) Beam trusses
(c) Beam truss cantilevers
(d) Space trusses
(e) Space truss cantilevers
(f) Structural connection fastener
(g) Edge of roof structurals
(h) Fascia structurals
(i) Panel fascia

5. Work not included:

(a) Anchor bolts
(b) All concrete, reinforcing bars, mesh, formwork and column-base grouting
(c) Attachment to walls and other nonsystem elements
(d) Roof and floor deck
(e) Roofing
(f) Miscellaneous iron and structural steel
(g) Air distribution devices and transition boots

GOVERNING STANDARD:

6. All work shall be performed in full accordance with current requirements of the __________ Building Code.

STRUCTURAL CALCULATIONS:

7. In addition to technical data, detail drawings and assembly drawings, the subsystem contractor shall submit complete structural calculations, signed and sealed by a professional engineer (structural) currently licensed to practice in the State of ________

8. Design of the structural system shall be in accordance with the latest editions of the AISC specifications for the Design, Fabrication and Erection of Structural Steel for buildings and the AISC specifications for the Design of Light Gauge, Cold Formed Steel Structural Members. Welding shall be in accordance with the American Welding Society for Welding in Building Construction. *(Continued)*

9. Data shall include column loadings, column uplift forces, and moments where bases are fixed or partially fixed.

10. Consideration shall be given to expansion and contraction and all solutions shall be related to the 5′ × 5′ planning module.

11. Loading conditions to be considered in the design of joints and connections are service loads, including wind, volume changes due to shrinkage, temperature change, erection loads, and loading encountered in storage and transportation of members.

12. The component structure shall be capable of supporting its self-weight plus the vertical loads, including uplift as indicated on the Structural Framing Plan and as required by code.

MATERIAL:

13. Steel shall conform to one or more of the current standard specifications of the American Society for Testing Materials.

- (a) Sheet and Strip Steel:
  - (1) A375 - low-alloy hot-rolled steel sheets and strips. Modified minimum yield point - 55,000 psi
  - (2) A375 - low-alloy hot-rolled steel sheets and strips. Minimum yield point 50,000 psi
  - (3) A245 - Grade-C flat-rolled carbon steel sheets of structural quality. Minimum yield point - 33,000 psi
  - (4) A446 - Grade-B zinc-coated (galvanized) steel sheets of structural quality, coils, and cut lengths. Minimum yield - 37,000 psi
- (b) Structural Steel:
  - (1) A7 "Structural Steel" minimum yield point - 33,000 psi
  - (2) A36 "Structural Steel" minimum yield point - 36,000 psi
  - (3) A529 "Structural Steel" minimum yield point - 42,000 psi
  - (4) A242 "High Strength Low Alloy Structural Steel" minimum yield point - 50,000 psi
- (c) Bolts, Nuts and Washers:
  - (1) A325 "High Strength Carbon Steel Bolts for Structural Joints Including Suitable Nuts and Plain Hardened Washers."
  - (2) A307 "Specification for Low-Carbon Steel Externally and Internally Threaded Standard Fasteners."
- (d) Welding shall be in accordance with the American Welding Society for Welding in Building Construction. Welding filler metal and flux shall fulfill requirements of Section 3 of the AWS Building Construction Code AWS D1.0-66.

FINISH:

14. All structural steel components shall be cleaned and prime painted. All primers used equal or exceed the performance requirements of TT-p-636.

SYSTEM DIMENSIONS:

15. Horizontal module shall be 5′ × 5′ planning grid for structure and ceiling. Depth of roof-ceiling sandwich shall be 35″ or less from underside of roof deck to ceiling grid surface. Floor-to-ceiling height shall be 12′. Columns shall measure 8″ × 8″ or less.

STRUCTURE:

16. Primary trusses shall be cambered and shall be designed to support loadings allowed on secondary members plus up to 5,000 lb load from mechanical rooftop unit centered on truss. Primary trusses shall allow the passage of ductwork in harmony with openings in secondary trusses.

17. Base members of secondary trusses shall provide integral 5′ × 5′ ceiling grid 3″ wide with 1″ wide slot between trusses for air supply and return.

18. The component structure shall be capable of meeting the following requirements:

(a) Superimposed Live Load	20 psf
(b) Dead Loads:	
Roofing and roof insulation	7.5 psf
Ceiling/lighting	3.5 psf
Sprinkler system and mechanical service lines	10 psf
Fascia line load	40 plf
Self-weight of structure	as required
(c) Wind Load	30 psf
(d) Uplift	30 psf

(e) The structural system shall be capable of resisting the superimposed wind load through column bending resistance; i.e., without the addition of any shear walls or cross bracing.

(f) Mechanical Equipment Load: As part of the work of this section, provide beams to support roof-mounted mechanical equipment.

ERECTION:

19. System contractor shall be responsible for the erection and the safe stability of the frame until the entire system is completed and capable of resisting all design forces.

20. Make adequate provisions for erection stresses and for sufficient temporary bracing to keep the system plumb and in true alignment until completion, including the roof deck and the elements which are part of the wind resisting system.

21. Erection operations, including the installation of temporary shoring, shall be carried out without loading portions of the system frame and nonsystem construction in excess of their safe load carrying capacity. Where the load capacity of the structure may be exceeded, shore constructed portions of the structure as required to positive foundation support.

22. Setting Base Plates: Set column bases on reinforced concrete foundation or piers true and level at the proper elevations. Use methods of vertical and horizontal adjustments during erection which permit full grouting.

23. Structural members may be stockpiled provided they have been previously inspected and accepted by such inspection agency as may be appointed by the Owner and provided storage is done in a manner which will not cause any damage to or will otherwise impair the strength or the finish of the members.

24. After erection, all welds, bolts, and abrasions in shop paint shall be painted with shop primer.

ENERGY CONSERVATION IN BUILDING DESIGN

Closely related to construction cost control is energy utilization control through building design. In the last fourth of the twentieth century there will be experienced irregular to continuing shortages of all sorts of fuels and sources. These will be accompanied by increasing incidents of electric power outages and voltage reductions. One hopes that this energy crisis will eventually be solved with new sources of energy, such as geothermal and solar energy, environmentally acceptable uses of higher-sulphur coal, or nuclear fusion. Almost certainly the solution will involve a combination of all of these.

However, even should the world's dwindling fossil fuels be managed in such a way and phased into new source development programs so as to avoid interim shortages, rationing, and outages (logistically feasible, but politically unlikely), the cost of all forms of energy will become a bigger and bigger part of owners' operating costs for buildings. In supplying energy for buildings, however, the factor of economy should be considered in relation to the nation's[6] and the world's interests in energy conservation.

The built environment (in the United States) uses about 33 percent of all the nation's energy. There are clearly opportunities to make energy demand reductions in the range of 20 to 30 percent in new buildings (as compared to typical buildings of the 1960s and early 1970s) without curtailing occupant life-style or significantly affecting construction cost in most cases. (In some cases energy conservation in building design may be accompanied by first-cost reductions.)

Therefore, it will be an irresponsible owner, architect, engineer, or construction manager who does not plan energy-conservative buildings to the extent feasible in each case.

The technology involved in making energy savings, in the range discussed above, is not particularly sophisticated and is surely within the reach of most practicing architects and engineers. In addition, there are a number of guides and publications available to architects and engineers on this subject. A very good one is a publication of the Center for Building Technology of the National Bureau of Standards entitled *Development of an Interim Standard for Energy Conservation in New Buildings*. Another is the U. S. General Services Administration's design guidelines prepared for GSA by the American Institute of Architect's nonprofit Research Corporation.

But, like so many other efforts at control of resources, energy conservation needs a management tool. Because the technology of conservation is not complicated and because the design and construction process *is* rather complicated, the most productive approach to energy conservation in buildings will be through a system of energy budgeting. What is needed is a tool that may be used by the management of architectural and engineering firms, construction managers, and owners. Without such a tool, efforts at energy conservation through the "laundry-list" approach would be like an owner

[6]With 6 percent of the earth's population, the United States, in 1973, used over 30 percent of all forms of the earth's available energy.

Lake Buena Vista Shopping Village, Lake Buena Vista, Florida. Lake Buena Vista Communities, Inc., Division of Walt Disney Enterprises, Inc. (Heery & Heery, Architects, Engineers, and Construction Managers.)

Hermetic Motors Plant, Small Motors Division, Westinghouse Electric Corporation, Athens, Tennessee. (Heery & Heery, Architects & Engineers.)

saying to his architect or construction manager, "Here is a list of ways to save money in construction which I require you to use, but I'm not going to tell you how much I can spend." It is difficult to imagine that sort of approach leading to realistic cost control. Likewise, the laundry-list approach alone will not lead to energy conservation unless it is put in the form of arbitrary codes in standards. The latter would be very unwise, as it will unnecessarily limit owners, architects, and engineers and disrupt segments of the construction industry. Just as energy budgeting provides a useful management tool, so does it provide the only intelligent approach to codes, statutes, and standards for energy conservation.

Following is an energy budgeting system which the author and his staff began developing in 1972:

ENERGY BUDGETING SYSTEM

Formulas are adopted for expressing all forms of energy in British thermal units (Btu's). In turn, criteria are adopted for a standard level of simultaneous operation of all systems (including vertical transportation, food service, and other equipment) within the building for (a) the heating cycle and (b) the cooling cycle.

Based on the building type and climatic region, a budget is established in terms of Btu's per hour per square foot for the heating and cooling cycles based on the standard criteria for a level of simultaneous operation of all systems. These standards are developed and checked out by three different methods:

1. A trial design using energy conservative subelements and checking resultant demand
2. Checking statistical data on existing buildings and making comparative judgments
3. Design analyses of existing buildings

Implementation is by design review with critical check on proper space classification, overall systems demand based on standard criteria for operation, and finally by limiting the size of energy service and storage, lead connections, main switches, and main valves.

Building permit issuance agencies could subsequently adopt a similar procedure, thereby bringing about a far more productive approach to energy conservation and at the same time allowing the individual owner and his architects and engineers a maximum latitude in design to fulfill the owner's individual requirements and desires.

In the absence of the required statistical data referred to above, architectural and engineering firms as well as construction managers who have the capability could develop their own data and guides.

Energy budgeting is a very useful service for the construction manager to provide. One productive approach for the separate construction manager would be as follows:

1. Develop energy budget at predesign phase along with cost budgets, programs, and schedules.
2. Make the energy budget part of the owner-architect/engineer agreement along with cost budgets, programs, and schedules.
3. Assist architect/engineer in evaluating methods of accomplishing the budgets.
4. Advise owner and prepare comments relative to architect/engineer's design submissions as to whether or not design meets energy budget requirements.

CHAPTER 7

A Cost-Control System

Design, industry responsiveness, estimating, and communications are the four keys to cost control. Inherently, they are interrelated.

Design and estimating must be a continuously combined process. Estimating and communications are inseparable. Communications between the owner and the architect/engineer or construction manager, between designer and estimator, and between designer and project manager are basic if there is to be the understanding that must precede cost control. Design and industry responsiveness, including the proper utilization of competition and what industry best produces, are so closely related that a project can fail both architecturally and economically from a lack of sufficient relationship in this one area alone.

Design and industry responsiveness are dealt with in other chapters. This chapter will deal with estimating primarily, communications as it relates to estimating, and the estimator/designer relationship. These are the major aspects of the cost-control component of the time/cost control design and construction management system.

If the architect/engineer has a combined design and construction management contract with the owner, the architect/engineer should take the responsibility of delivering to the owner the construction contract or contracts that are within the budget previously established.

If, instead, there is a separate construction manager, that firm should take the same responsibility. This makes clear that the construction manager must be on the scene at the outset and must have that degree of control that has been discussed in previous chapters.

Setting the budget or economic requirements is, of course, fundamental. It is almost as bad to set budgets too high as it is to set them too low. It is certainly the easier course for the project that does not have to stand the test of economic feasibility. But setting the budget too low for the commercial project in order to have the appearance of an acceptable *pro forma* can be disastrous.

Obviously, the owner must play a major role in setting the budget or approving it. But most owners are not equipped to set a budget without help. And it is a cruel and unprofessional game for the architect/engineer to take the position that he must be handed the budget and the program before he can start work. It is true that budgeting and programming are extra services not covered under most architectural and engineering contracts, but once the architect/engineer or construction manager has made necessary arrangements with the owner for these services, that firm and the owner should work together in establishing a realistic program and budget.

The foregoing would normally take place before design work starts. However, there are projects for which it is not feasible to set a realistic budget until some design work has been accomplished. But it is the rare project that should not have a firm budget set by the time the schematic design work is complete, which would normally consist of about 15 percent of the architect/engineer's services.

Most projects, though, can have realistic maximum budgets set before design work starts. This would certainly be true of simpler new-buildings projects on uncomplicated sites. Normally this would include the majority of schools, apartments, hospitals, motels, parking garages, industrial plants and warehouses, office buildings, dormitories, classroom buildings, and other projects that are straightforward or for which there is sufficient comparative data available.

Most commercial projects will have their budgets set by the economic requirements. A *pro forma* that shows projected income and operating expenses along with debt service and return on investment determines what may be invested in land, design, construction, interim financing, and other costs of improvement and development. In turn, the lower the cost of the construction within acceptable quality levels and design goals for the maximizing of income, the better the job the architect/engineer or construction manager has done for the owner.

Once the construction cost limitation has been set in the project, the appropriate amounts should be deducted from it for the in-progress contingencies and cost escalations during the projected preconstruction period. The residual sum, which would constitute the maximum allowable construction contract (or contracts) award price, should then be related to the income projects by income units. In an office building, this would normally be arrived at by the relationships of net rentable floor area to gross floor area to construction contract award price, with cost of space subdivision and tenant fit-up taken into account.

In an apartment project, it will be the number of rentable or salable dwelling units of the respective types and requirements.

In a parking garage, it will be the number of parking spaces.

In a shopping center, it will be the number of rentable square feet of store area.

In a hotel or motel, it will be the number of guest rooms and other income units such as restaurants, meeting rooms, exhibit halls, etc.

These units should then be compared with applicable statistical data, or they may be tentatively set in this manner first in order to have a beginning point for running a *pro forma* test.

Applicable comparative data is often difficult to assemble and maintain in usable form. It can also be very easily misused. Assembling this type of data that is reliable, maintaining it continuously with updating information, and making it available in usable and variable forms is a fundamental responsibility of the estimator or the cost-control department of the construction management or architectural and engineering firm.

The main pitfalls in using comparative data are

1. Difficulties in setting up data and project units on the basis of direct comparables.

2. The tendency to accept the average or previous accomplishments as acceptable goals and, in turn, failure to look for cost reduction or quality increase through better design or different approaches.

The latter may seem idealistic, but nonetheless the architect/engineer and construction manager must always retain his objectivity and keep the owner's best interests foremost in his considerations.

As to the former, difficulties in setting up comparative data can be overcome by careful attention to the following:

1. Standardizing methods of calculating area and volume.

2. Identifying time and location factors through the use of a good cost index such as the *Engineering News-Record* Building Cost Index.[7]

3. Identifying site and subsurface conditions variables.

4. Identifying site-work variables.

5. Identifying equipment and furnishing inclusion variables.

6. Identifying significant mechanical/electrical requirements variables.

7. Identifying nonconstruction inclusion variables—such as design, financing, and land costs—and omitting these.

8. Identifying any ancillary facility prorations. (For example, in a hotel project, some analysts will include pro rata costs of swimming pools, restaurants, lobbies, meeting rooms, maintenance shops, etc., in the cost of the guest rooms. Others may include some but not all of these types of facilities. Still others may include none. A standard format must be established for each building type in the comparative data. However, it is often needed by designers, project managers and owners in several different formats of ancillary facility costs prorations for the purposes of running different comparative checks.)

A program of comparative statistical construction cost data by building types is obviously one area of application of electronic computers in the practice of architecture and engineering as well as construction management that has good potential.

In projects other than commercial projects, the comparative data method is also most useful in setting budgets. This would certainly apply to schools, hospitals, dormitories, public housing, industrial plants, and many others.

In many projects, it is more practical to convert units other than floor area units to gross floor areas. For example, in a school, budget criteria would likely

[7]Appears weekly in *Engineering News-Record,* a publication of McGraw-Hill, Inc.

be number of students; but most states will have guides or regulations for converting student population to square footage. Industrial plants have their basic economic justification in production or storage capacities; however, these are quite frequently converted to gross floor areas of the different types of spaces called for in the program.

Following are some of the building types that may well remain in unit form other than square feet into the design phase:

Apartments and housing (units)
Hotels and motels (guest rooms)
Parking garages (spaces)

Other than the comparative data method, there are three other ways in which budgets may be set. They are

1. Determination by funds available, in which case the program becomes the variable, being set by utilizing the comparative method above or by one of the following two methods. This is frequently called working back from the amount available.

2. Determination by preliminary design.

3. Determination by predesign analysis. This approach usually consists of a mixture of the comparative method and 2 above. For example, if the project consists of a campus area with several buildings of different types, drives, parking, and landscaping and is accompanied by unusual subsurface conditions, the budget may be prepared as follows: The various buildings are budgeted on a comparative unit basis. This could be by floor areas for classroom and administrative buildings and by units for dormitories. To these building figures are added sums for the subsurface conditions. This information would have to come from experienced structural engineers and be based on soil tests and building height assumptions. In turn, an experienced estimator should be able to set up budgets for the site work based on a review of the topography, the soil tests, number of cars to be parked, site development program requirements, and discussions with the owner or designers.

Once the figure for the construction contract(s) award price(s) is determined, it should be entered in the project budget. This should be done as early as possible. Further, a standard format should be established for the project budget at the outset and the same format should always be used thereafter without fail. It is a most important part of the communications aspect of cost control that this be done. Such a format should always show all related items and should include provisions for in-progress contingencies and cost index adjustment from the outset.

Figure 7-1 illustrates a simplified recommended project budget form. Note three significant terms in the form that will be used below. They are: Construction Contract(s) Award Price(s) (CCAP), Estimated Total Construction Cost, and Total Project Budget.

FIGURE 7-1

RECOMMENDED PROJECT BUDGET FORM

EXHIBIT A

For Agreement between ______________________________, THE OWNER,
and HEERY & HEERY, INC., dated ______________________, 19 ________
Comm. No. ________

____

____

PROJECT BUDGET

1. This budget is based on construction dates and times as follows:

2. Escalation rates shall be determined by the ______________________ Cost Index for ______________________ (city). Based on recent ______________________ ______________ Index history in the area, an escalation of ______________% has been allowed for in the figures below based on the above dates and times. *In the event of delay, all of the following figures are automatically adjusted by the foregoing Index.*

3. Budget for Construction Contract(s) Award Price(s), herein referred to as CCAP: (List proposed separate construction and purchase contracts, if more than one, including interiors and graphics.)

Total CCAP $ ....

4. In-progress contingency fund to cover change orders for necessary adjustments to site conditions, minor design refinements, and correction of minor errors and omissions in the construction documents.
______________ percent (%) of Item 3 above. $ ....

NOTE: The above fund is not for changes or additions to the Project. If the Owner desires a fund to cover such contingencies, said fund should be set up in addition to the foregoing. (See Article 3 of the Contract between Owner and H&H with regard to increases in the Project Budget.) ______________________

5. Total Project Construction Budget $ ______________________

(Continued)

6. Recommended Allowances for other items:
   (a) Architectural and Engineering Services $ ....

   (b) Construction Management Services $ ....

   (c) Consultants:
   CPM ________________________ $ ....
   Acoustics ________________________ $ ....
   Traffic ________________________ $ ....
   Food ________________________ $ ....
   ________________________ $ ....
   ________________________ $ ....

   (d) Additional Services
   Models and Renderings ________________________ $ ....
   ________________________ $ ....
   ________________________ $ ....

   (e) Full-time Construction Management Personnel, time; $ ....
   travel and living. $ ....

   (f) Reimbursables under Owner-H&H Contract, travel: $ ....
   reproduction; $ ....
   Other. $ ....

   (g) Surveys, Tests, Borings, Reports $ ....

   (h) Other:

Total Item 6 ________________ $ ________________

7. Total Project Budget (ITEM 5 PLUS ITEM 6) $ ________________

NOTE: If this Exhibit form is completed and approved simultaneously with the execution of the contract or prior to the submission of Schematic Phase Documents to the Owner, there shall be attached hereto a Program of Requirements which will describe and list the Owner's functional requirements for the Project on which the Project Budget is based.

With the budget established and all parties in agreement on it, the design work and the cost control system's implementation start together.

At this point, it should be recognized that there is no magic, no secret formula, no computer program, or even a set of stereotyped procedures that will ensure cost control. Possibly the procedures discussed below should not even be called a system because the term "system" may tend to lull designers and others into a false feeling of security. The fact is that cost is controlled through continuous care and thoroughness backed up with construction and estimating experience. The system consists primarily of a group of procedures that are simply a part of a continuing close relationship between design, estimating, and project management.

Fundamental to implementing the cost control system is the estimator or cost control group of the architect/engineer or construction manager.

It is not realistic for an architect/engineer to undertake a design and construction management service or for a construction manager to undertake to provide separate construction management services without at least one good, in-house, construction-experienced estimator. The cost-control staff size should be proportionate to the size of the firm and its projects. There must be sufficient estimators to keep up with all ongoing projects as well as all budgeting and data compilation.

The estimator or cost control group must be capable of doing the following:

1. Comparative data compilation and presentation.
2. Preparation of budgets without benefit of designs.
3. Participation in and contribution to the predesign project analysis including budget analysis, construction sequencing analysis, and identification of projected labor work stoppages and wage increases.
4. Distributing budget into building design components. For example, the composite exterior skin. Other examples would be the plumbing system as a total, prorated into fixtures and features, or the structural frame above the foundations on an area or volume unit basis.
5. Estimating by units.
6. Take off and estimating in detail by labor and material.

After the predesign project analysis, the first activity of the estimator should be to provide the designer with the following:

Areas of potential cost savings
Areas of potential cost problems
Project's components budgets

The cost control system described here consists of two parallel but independent estimating programs maintained throughout all phases of the design work. One is conventional unit price and labor/material estimating. The other is budgeting of components, as mentioned in 4 and 6 above, with continuous rechecks as design evolves.

In the normal course of architectural and engineering services, the design work and certain documentation that includes estimates is submitted for the owner's review and approval. The traditional architectural contract has taken the position that the purpose of the several estimate submissions to the owner at these phases is to advise the owner of the current estimate and as to whether or not there has been any change in the estimated costs from the previous estimate. This is not an acceptable philosophy of good construction program management, whether provided by the architect/engineer or a separate construction manager.

Only in the most unusual projects, small alteration projects, or in the most volatile and unpredictable of construction markets is the above an acceptable approach for most owners. The continuing estimating is done for the sake of continuous design check, adherence, and, as necessary, corrective action within the program of requirements. Estimate submission at the various phases should be done only for the sake of verification and the owner's accounting purposes plus the architect/engineer's or construction manager's internal purposes.

When an estimator says to the designers or project manager that he can work only from fairly detailed drawings and specifications, the time has come to get a new estimator. That man is only a takeoff and pricing technician, not truly the estimator that project managers and designers must have available. At the same time, the project managers and designers must understand the basis of estimates made from less than complete drawings and specifications. When an estimate is made on that basis, it must be recognized to be a prediction or design budget. The latter is usually the case. In other words, the estimator, who should be in close communication with the designers, is budgeting sums within which it would be reasonable to expect the various components of the project to be designed. Of course, this budget must be based on a good understanding of what the designer had in mind and how he is likely to execute it.

The latter situation further points up the real need for the estimator to be in-house or, as a maximum compromise, to be made available continuously by the separate construction manager to work with the architect/engineer's office.

The best estimator/designer teams will be those who develop the experience of working together over a reasonable period of time, so that each learns what to expect from the other.

In preparing preliminary estimates or budgets from less than complete drawings and specifications, a good procedure to follow is to be sure that at least some reasonable figure is in the estimate for each item that is likely to be in the project, whether or not it appears in the early sketches or program of requirements. One good way to do this, of course, is to employ an estimator's check list. Figure 7-2 illustrates such a check list.

FIGURE 7-2

ESTIMATOR'S CHECKLIST
HEERY & HEERY, ARCHITECTS & ENGINEERS

CSI Division

1. General conditions:
   - Temporary facilities
   - Bonds & insurance
   - Supervision
   - Temporary utilities

2. Site work:
   - Demolition
   - Remove debris
   - Clear & grub
   - Grading
   - Paving
   - Site improvements
   - Special foundations

3. Concrete:
   - Material and placing by type (footing, wall, slab, etc.)
   - Formwork
   - Reinforcing
   - Accessories
   - Finishing
   - Precast

5. Metals:
   - Structural steel
   - Miscellaneous steel
   - Metal decking
   - Metal specialties

6. Carpentry:
   - Rough carpentry
   - Wood blocking
   - Millwork
   - Finish carpentry
   - Flooring & paneling

7. Thermal & moisture protection:
   - Roofing
   - Insulation
   - Wall panels
   - Waterproofing
   - Fireproofing
   - Special protective items

8. Doors and windows:
   - Doors (by type)
   - Frames (by type)
   - Rated doors and frames
   - Window wall
   - Storefront
   - Glazing
   - Special doors

9. Finishes:
   - Floor covering
   - Ceiling construction
   - Partition construction
   - Wall covering (paint, tile, vinyl, etc.)
   - Graphics

10. Specialties:
    - Toilet partitions, urinal screens, etc.
    - Bulletin boards, chalk boards, etc.
    - Miscellaneous specialties

11. Equipment:
    - Kitchen equipment
    - Loading dock equipment
    - Commercial equipment
    - Laboratory equipment
    - Medical equipment
    - Detention equipment

12. Furnishings

13. Special construction:
    - Swimming pool
    - Sound- or vibration-free rooms
    - Clean rooms
    - Prefabricated buildings or rooms

14. Conveying systems:
    - Elevators
    - Escalators
    - Lift systems

15. Mechanical:
    - HVAC
    - Plumbing
    - Waste treatment
    - Fire protection

16. Electrical:
    - Power systems
    - Special systems
    - Communication systems

By the time the architect/engineer has completed the schematic design phase, either the in-house estimator or the construction manager should have completed the schematic phase reconfirmation of the CCAP and the full project budget as shown in Figure 7-1. While it is good practice always to make an estimate or reconfirmation in as much detail as is feasible, detailed quantity surveys may not always be feasible at this design phase. At a minimum, however, the estimate at this phase should be on a square-foot or volume-unit basis plus items that do not lend themselves to this type of unit estimating. Figure 7-3 illustrates such a breakdown of the CCAP for a project. This document would be the backup for the project budget submitted for reconfirmation at the schematic design phase.

FIGURE 7-3

ESTIMATE SUMMARY SHEET

XYZ CORPORATION
MANUFACTURING FACILITY

COMM: 7314

ESTIMATOR: CAGIII,DOK,RMG

DATE: August 7, 1973

PHASE: Schematic

SHEET 1 OF 1

ITEM	DESCRIPTION	COST	CHK.
2	SITE WORK	215000	
	FOUNDATIONS	16000	
3–10	WAREHOUSE AREA 46000 SF @ 5.50	253000	
3–10	MANUFACTURING AREA 56000 SF @ 5.60	313500	
3–10	OFFICE AREA 8000 SF @ 17.50	140000	
	SUBTOTAL	937500	
	20% OVERHEAD & PROFIT	187500	
	SUBTOTAL 2 THRU 10	1125000	
15	MECHANICAL SYSTEMS FOR WAREHOUSE AREA @ 3.75	172500	
15	MECHANICAL SYSTEMS FOR MANUFACTURING AREA @ 5.70	319200	
15	MECHANICAL SYSTEMS FOR OFFICE AREA @ 6.00	48000	
16	ELECTRICAL FOR WAREHOUSE AREA @ 2.00	92000	
16	ELECTRICAL FOR MANUFACTURING AREA @ 2.25	126000	
16	ELECTRICAL FOR OFFICE AREA @ 2.50/SF	20000	
	SITE UTILITIES	17750	
	SUBTOTAL	795450	
	10% OVERHEAD & PROFIT	79550	
	SUBTOTAL 15 AND 16	875000	
	5% DESIGN CONTINGENCY	100000	
	TOTAL CCAP	2100000	
	3% IN-PROGRESS CONTINGENCY	63000	
	TOTAL CONSTRUCTION BUDGET	2163000	

As will frequently be the case, at various times after the original project budget has been set, the owner will decide to make program or design modifications that will have an effect on the cost. Sometimes these decisions will come as a result of recommendations from the architect/engineer or construction manager. It is in this area that a large number of so-called cost problems occur. It is difficult to say who has historically been more at fault, the owner for failing to recognize that such changes affect the cost or the architect/engineer for failing to communicate this to the owner with adequate documentation in the project files, confirmed in writing by the owner.

Figure 7-4 illustrates a recommended form for a design change order. Note that this form is similar to the construction contract change order, so that all

FIGURE 7-4

HEERY ASSOCIATES, INC.

880 W. PEACHTREE ST., N.W., ATLANTA, GEORGIA 30309 (404) 881-1666 Telex 54-2165

Design Change Order

FOR CHANGES TO THE SCOPE OF THE PROJECT

PROJECT:

DATE:

D.C.O. No.

The change hereinafter described is applicable to and is hereby made a part of the design for the above referenced project.

Description of Change:

Exhibits ____________ Dated ____________ shall be and are hereby made a part of this change order.

This change will ______________ the total project construction budget in the amount of $____________
increase or decrease
and will affect the schedule as follows:

RECAP OF PROJECT CONSTRUCTION BUDGET

BUDGET	ORIGINAL	PRIOR ADDITION OR DEDUCTION	THIS CHANGE ADDITION OR DEDUCTION	CURRENT AS REVISED
Construction Contract Award Price Plus In-Progress Contingency				
Total Project Construction Budget				

Recommended for acceptance:

________________________________ ARCHITECT

By ____________ Date ____________

HEERY ASSOCIATES, INC. CONSTRUCTION MGR.

By ____________ Date ____________

Approved and agreed to by owner: ____________

By ____________ Date ____________

parties recognize its significance. Also, the reader should note that all design change orders to date should be recapitulated in each submission of the project budget (Fig. 7-1). This procedure should be followed for the work under each contract up until the construction contract is awarded.

By the time the architect/engineer has completed the design development phase of his services for the owner (usually around 35 percent of services), the in-house estimator or the construction manager should have the CCAP of the project budget reconfirmed with a detailed quantity survey estimate. Figure 7-5 illustrates a typical form for such an estimate.

During the design development work of the architect/engineer and in the contract document preparation phase, as in the schematic phase, there must be a continuing dialogue between designers and estimators. Once the design development documents are approved by the owner and the architect/engineer has been authorized to commence the construction contract documents (final working drawings and specifications), the first activity of the estimator should be a recheck or complete quantity survey reestimate of the design development documents, taking into account any modifications or adjustments as a result of this final preliminary design review. And, as the contract document preparation proceeds, the estimate should be redone or rechecked in detail again so that the project manager and the designers will have a final estimate sufficiently in advance of bidding or negotiation to make necessary final adjustments. This last estimate should include estimates of all planned alternate price items and any alternates recommended by the estimator.

Having come to the completion of the contract documents for a project or a component of a project, the factor that will mean that estimates and predictions will be reliable is industry responsiveness. For group I owners, this will mean at least adequate competition. For group II owners, this will usually mean the owner's relationship with the selected contractor. However, in the latter case, the architect/engineer or construction manager must remember that the contract documents, in turn, should maintain the contractor's competitive position with material suppliers, manufacturers, and subcontractors. Also, the project procedures and document format for group II owners should always be such than an alternate course (bidding rather than negotiation) could be followed if necessary.

For group I owners, sufficient competition is essential and is covered in some detail in Chapter 11, as are some pointers on direct negotiation for group II owners.

During the construction phase, the major cost-control activities are eliminating change orders and fully enforcing the contract through the proper handling of progress and final payments. However, it would be naïve to believe that any construction contract will run its course without some change

FIGURE 7-5

ESTIMATE SUMMARY SHEET

XYZ CORPORATION
MANUFACTURING FACILITY
ESTIMATOR: CAGIII,DOK,RMG

COMM: 7314
DATE: October 5, 1973

PHASE: Final

SHEET 1 OF 1

ITEM	DESCRIPTION	COST	CHK.
2	SITE WORK	225047	
3	CONCRETE	285395	
4	MASONRY	47072	
5	METALS	62995	
6	CARPENTRY	10125	
7	THERMAL AND MOISTURE PROTECTION	83070	
8	DOORS & WINDOWS	54315	
9	FINISHES	139009	
10	SPECIALTIES	31347	
	SUBTOTAL	938375	
	20% OVERHEAD & PROFIT	187675	
	SUBTOTAL	1126050	
14	HVAC	368100	
14	PLUMBING	134300	
14	FIRE PROTECTION	76420	
15	ELECTRICAL	291130	
	SITE UTILITIES	17106	
	SUBTOTAL	887056	
	10% OVERHEAD & PROFIT	88706	
	SUBTOTAL UTILITIES	975762	
	TOTAL ESTIMATED CONSTRUCTION CONTRACT AWARD PRICE	2101812	
	3% IN-PROGRESS CONTINGENCY	63054	
	TOTAL CONSTRUCTION BUDGET	2164866	

orders, and the owner cannot reasonably expect any set of contract documents to be perfect. Therefore, one must be prepared, contractually and otherwise, for change orders.

Most construction contracts will provide, in their general conditions, for various methods of pricing change orders. These usually include the following:

1. Negotiated price
2. Unit price in the contract
3. Negotiated unit price
4. Accounting of labor and materials after the work has been completed

Figure 7-6 is a recommended contract provision to be used in conjunction with the AIA standard general conditions relative to method 4 above.

FIGURE 7-6

SPECIMEN SECTION ON CHANGE ORDER MARK-UP LIMITS

(Supplementary General Conditions)

Add the following subparagraphs to paragraph 12.1.3:

In sub-paragraphs .1 and .3, the allowance for overhead and profit combined, included in the total cost to the Owner shall be based upon the following schedule:

For the contractor, for any work performed by his own forces, 15% of the cost;

For each subcontractor involved, work performed by his own forces, 15% of the cost;

For the contractor, for work performed by subcontractor, 5% of the subcontractor's cost plus subcontractor's percentage fee as defined above.

Cost shall be limited to the following: Cost of materials, including sales tax and cost of delivery, Cost of Labor, including Social Security, Old Age and Unemployment Insurance (Labor cost may include a pro rata share of superintendent's time, only in case an extension of Contract Time is granted on account of the change); Rental of power tools and equipment (including fuel).

Overhead shall include the following: Bond premiums, supervision, superintendence, wages of timekeepers, watchmen and clerks, small tools, incidentals, general office expense. Workmen's Compensation Insurance, and all other expenses not included in "Cost."

At the end of the second sentence in paragraph 12.1.4 delete the words "a reasonable allowance for overhead and profit" and substitue the following in lieu thereof:

"an allowance for overhead and profit in accordance with the Schedule set forth above in Paragraph 12.1.3".

Unit prices can easily be misused or employed when they should not be. It is common practice for the proposal forms for lump-sum general contracts to also have unit-price quotation blanks for such things as rock removal. This is generally a very unwise procedure because, in the lump-sum contract, the award is based on the low bid and very little attention is or can be given to these unit prices. Therefore it is better, usually, to include specified allowable unit prices for additions to and deductions from the contract in the specifications. Another approach that may be more feasible in many cases is to have rock removal, extra excavation, etc., done under method 4 above, which is sometimes referred to as the "force accounting" method. In this case, the architect/engineer or construction manager can prepare a rough estimate or general description of the work for the owner before it proceeds, but because of the nature of this type of unknown work, one should make it very clear that there is often no way of making a reliable estimate for the cost of the work. Negotiated unit prices is another approach, but this is generally less reasonable for the owner than the former two methods.

There are some kinds of items for which unit prices must be obtained in the lump-sum contract proposal. An example is foundation piling, since the lengths and quantity can almost never be set exactly except by the pile driving itself. There is little excuse, though, for not having a good estimate of piling lengths and capacities based on soil tests and test piles. Based on these, the lengths of the piles (and bearing capacity ratings if there are differently rated piles planned on the project) should be set as closely as the engineers can estimate,with a very slight margin of safety on the high side. In turn, the unit price called for in the proposal should be a figure that is specified to be applicable for both additive and deductive contract price adjustment based on actual pile lengths.

Many owners have unnecessarily cost themselves a greal deal of money in change orders by requiring their architect/engineer or construction manager to obtain firm quotations for all change orders in advance of the approval of the proposed additional work. Unnecessary extra cost is usually the case when the need for a change order develops and it is clear to all concerned, including the contractor, that the work in question must be undertaken and that the owner is requiring a firm quotation in advance. In this event, the owner is in an untenable position, and it is not unusual in these cases for the change order to cost from 200 percent to 500 percent of what it should cost. It is true that labor and material accounting is difficult at best, and it is often true that the owner does not get work done under change order for prices as reasonable as those he receives in the main contract proposal. But nonetheless, when it is clear to all concerned that there is no way to avoid the change order work, the force accounting method will usually be best except in the cases of the type of unit prices described above for pilings.

In preparing the project budget, an in-progress or "construction" contingency allowance must always be included. This amount plus the cost index adjustment and the CCAP gives the estimated total construction cost. The contingency allowance should be for those things that are listed for it in Figure 7-1, it not being reasonable for such an allowance to be set up to cover all instances of the owner making changes and additions to the project. There need to be, in turn, very clear communications with the owner about this item from the outset. The owner must have this as well as the other figures in the budget fully explained, preferably in writing. It should be pointed out at the predesign setting of the budget and reiterated at the execution of the construction contract what the purposes of this sum are. The owner should be cautioned both times to make additional allowances for any sum the owner feels is necessary for his in-progress changes or additions.

Using the in-progress cost controls described above, change orders for site condition adjustments, minor changes by the architect/engineer, and corrections have run between 1 to 1½ percent for most projects. However, for setting the budget, around 3 percent of the CCAP is usually a reasonable sum.

An owner should be able to expect as jobsite adjustments, minor changes recommended by the architect/engineer or construction manager, and corrections of minor errors in the contract documents (excluding alterations, additions, and remodeling) to fall within a 3 percent allowance for most projects except those that are very small. Conversely, the owner has no reasonable complaint about the contract documents or the handling of the project if the figure falls within that range or whatever figure is agreed to in advance by the owner, architect/engineer, and construction manager.

The figure of 5 percent has been commonly used by some for in-progress contingency allowances, but this would seem excessive unless controls like those described in this chapter are ignored, unless there is going to be a decentralization of contract administration, or unless the 2 percent were pulled out of the air and added for the owner's changes of mind and additions.

In conclusion, a well-known football coach once told his players, "There is only one defense: guts." In cost control of construction, there is only one system: tenacity.

CHAPTER 8

Time Control, Contract Time Provisions, and Contract Time-Extension Rulings

Time control and project acceleration will best be accomplished in building design and construction by the following basic project management actions:

1. Analyzing the owner's posture for purchasing construction relative to the responsiveness of contractors and other elements in the construction industry, and obtaining bids for or negotiating construction contracts thereafter accordingly.

2. Identifying all constraints relative to the site, design, construction, owner, and required approvals.

3. Scheduling all design and construction activities, along with all owner and third-party activities that are critical requirements, based on the identified constraints and the desired occupancy dates or earliest feasible date depending on the owner's needs. Contingency time for likely extensions, including identified probable labor work stoppages, should be allowed.

4. Awarding early any construction contracts identified as constraints (sometimes called "phasing"), all the while working toward having, if possible, a single-responsibility general contract by the time general work commences at the site. If the single-responsibility general contract is not feasible or allowable, it is best to come as close to having one as is possible in the individual situation, realizing that alternative approaches occasionally, in certain special situations, may be in the best interest of the owner. [Alternative approaches referred to include separate major contracts (prime contracts ranging from three to eight for the typical project) and separate trade contracts (ranging in number from thirty to forty for the typical project). In the latter case, there would be no general contractor, and the construction manager (whether architect/engineer or separate manager) would assume the supervision and coordination responsibilities of the general contractor.]

5. Avoiding unnecessary phasing. Having drawings and specifications complete for each awarded contract. When early bidding is indicated by the constraint identifications, avoiding early contract awards whenever feasible and planning the foregoing procedure in the contract documents in advance.

6. Employing construction agreements and contract documents that are carefully tailored to the individual project and owner. Reexamining especially the general and special conditions of the contract documents and making the fullest use applicable, in each case, of liquidated damages provisions, progress payment scheduling, superintendent controls, change order rights for the owner, occupancy rights for the owner, clearer and fewer contract time extension provisions, and scheduling requirements.

7. Being sure there is adequate competition whenever bidding is employed. Being sure that bidders thoroughly understand and have scheduled their work in advance of placing proposals.

8. Centralizing contract administration in the construction manager

(whether architect/engineer or separate manager). Not allowing those involved in contract administration or field services to proceed with their work without a thorough familiarity with the time-control provisions of the contract and the schedule.

9. Being diligent and resourceful throughout. Not placing reliance on any one or two techniques such as computer-assisted scheduling, industrialized building systems, or phased construction. Instead, doing everything that can be done for each project without concern for management system redundancy.

The foregoing not only constitutes a series of recommended actions but also embodies the basic philosophy of the time/cost control system insofar as time control and project acceleration are concerned.

The constraints in a project, its design, its schedule, and its construction management plan will determine feasibility of scheduled beneficial occupancy and final completion dates. The provisions of the contract, and its administration, will in large measure determine the dependability of those dates for group I owners. Responsiveness by contractors, because of the prospect of repeat business, may be a bigger factor for time-of-completion dependability for group II owners. However, the same provisions, or similar but less formal ones, are still well worth having in construction contracts for group II owners.

The following are discussions of a number of time-control contract provisions for which specimen provisions are contained in this chapter. The specimen provisions are for group I owners, but they could be used substantially as illustrated in almost any construction contract. It should be pointed out that these are tried and proved provisions, having been developed through usage, for both private and governmental owners, for a wide range of building types and sizes in many parts of the United States over a period of some fifteen years.

In instances of there being more than one contract for a project—such as separate contracts for foundations, mechanical work, systems, etc.—the same provisions that are discussed below, with appropriate minor wording changes, should be included in all contracts. This would apply whether or not the separate contract is to be transferred into a general contract.

Contract Time Extensions (Figure 8-1)

The typical construction contract, as may be seen by reading the American Institute of Architects' standard general conditions is of necessity written for general rather than specific situations. Further, it is often the result of a series of compromise agreements between various industry associations and professional organizations in sessions that do not have owner representation. Consequently, when the specifics of the individual situation and the owner's best interests are given first consideration, a better contract can usually be obtained for the owner. This is true not only for contract time-extension provisions but also for a number of other aspects of the construction contract.

Figure 8-1 would constitute a replacement for the AIA corresponding general conditions article or similar articles of other construction contracts' general conditions.

It should be noted that this provision, as well as the other time-control provisions illustrated herein, is predicated on there being stipulated calendar dates or elapsed calendar days for beneficial occupancy and final completion dates in the construction contract proposals. One should never use a proposal form that leaves the time of completion up to the bidder to fill in as part of his proposal placement. (Nor should one ever use "working day" scheduling.) If there is some doubt about the feasibility or extra cost of a particularly tight schedule, alternate bid prices should be included in the proposal form for the specific alternate dates or contract times, with all bidders proposing on the same base bid and alternate dates. Throughout specimen section, substitute "Construction Manager" for "Architect" in most cases, if a separate CM is handling or coordinating the construction phase.

FIGURE 8-1

SPECIMEN SECTION ON CONTRACT TIME EXTENSION

(Supplementary General Conditions)

Delete paragraph 8.3.1 of the General Conditions and substitute the following in lieu thereof.

There will be no extensions of the Contract Time except under one or more of the circumstances listed below. Such extensions shall be granted for such reasonable time as the Architect may determine. In ruling on claims for extension in the Contract Time, the Architect may take into account whether or not a given occurrence should, in fact, have any effect on the Contractor's ability to meet scheduled dates of Substantial Completion (Beneficial Occupancy) or Final Completion.

a. Delay in progress due to an act of neglect by the Owner or the Architect or delay in work by others specifically listed in the Schedules and Reports Section.
b. A Change Order approved by the Owner that contains an extension of the Contract Time approved by the Architect. However, the Contractor agrees that extensions in Contract Time granted in Change Orders are subject to extension-of-time audit by the CPM Monitor.
c. Labor strikes, beyond the control of the Contractor (including strikes affecting transportation), that do, in fact, directly and critically affect the progress of the work. However, an extension of Contract Time on account of an individual labor strike shall not exceed the number of calendar days of work stoppage resulting from that strike.

(Continued)

d. Tornado, hurricane, lightning, blizzard, earthquake, typhoon or flood that substantially damages completed work or stored materials provided that an act of neglect of the Contractor did not contribute to such damage. Except for these circumstances, there will be no extension in the Contract Time due to normally inclement weather, unless the Contractor can substantiate that there was greater than normal inclement weather considering the full term of the Contract Time using accumulated record mean values from climatological data compiled by the U.S. Dept. of Commerce National Oceanic and Atmospheric Administration for the locale.

Extensions in the Contract Time in Change Orders are subject to extension-in-time audit by the CPM Monitor as follows:

(1) The Contractor agrees that, even though the Owner, Contractor and Architect have previously signed a Change Order containing an extension-in-time resulting from a change in or addition to the work (not as a result of a through d above), that said extension in the Contract Time may be adjusted by an audit after the fact by the CPM Monitor. If such an audit is to be made, the CPM Monitor must undertake the audit and make a ruling within 60 days after the completion of the work under the change order.

(2) The Contractor agrees that any extension of the Contract Time to which he is entitled arising out of a change order undertaken on a force accounting (labor and materials) basis, shall be determined by an extension-in-time audit by the CPM Monitor after the work of the change order is completed. Such rulings shall be made by the CPM Monitor within 60 days after a request for same is made by the Contractor or Architect, except said 60 days will not start until the work under the change order is completed.

Rulings of the Architect and CPM Monitor on extensions in the Contract Time are exempted from any mandatory arbitration provisions of this Contract. Should the Contractor or Owner wish to challenge an extension in Contract Time ruling of the Architect or the CPM Monitor, the following procedure shall be followed, and the resulting appeal ruling shall be binding on both parties:

Upon receipt of written demand for appeal by the Contractor or the Owner, the Architect shall empanel the following review parties:

a. The Architect
b. The CPM Monitor
c. A party appointed by the Contractor

Both parties shall then submit the facts in writing plus their respective requested ruling.

The above listed panel shall promptly convene, may waive all formal procedures of arbitration and legal proceedings, may waive appearances by both parties (not one party only), and shall submit the new ruling or reconfirm the previous ruling.

If the previous ruling is reconfirmed, the party requesting the appeal shall pay all expenses of the review, including time charges from the panel members.

If the appealed Contract Time extension is adjusted, the Owner and the Contractor shall equally divide the above described expenses of the review.

Definition of Beneficial Occupancy and Substantial Completion (Figure 8-2)

This illustrated provision is supplementary to the AIA general conditions. It will be noted that the terms "beneficial occupancy" and "substantial completion" have the same definition, which may not be completely logical in the minds of attorneys. The reason for this is that "substantial completion" is an industry term that has historically been somewhat ill-defined until some of the more recent editions of the AIA general conditions. Experience with the time/cost control system has shown that it is best to reinforce the more complete definition of the term with the descriptive term "beneficial occupancy," which is used throughout the time/cost control system in both schedules and in contracts.

FIGURE 8-2

DEFINITION OF SUBSTANTIAL COMPLETION AND BENEFICIAL OCCUPANCY

(Supplementary General Conditions)

The term "Beneficial Occupancy" is interchangeable with the term "Substantial Completion." A state of beneficial occupancy will be reached when all work is complete, accessible, operable and usable by the Owner; all parts, systems and site work is 100% complete and cleaned for the Owner's full use. Only incidental corrective work under "Punch lists" and final cleaning (if required) beyond cleaning for Owner's full use may remain for Final Completion.

Liquidated Damages Provisions (Figure 8-3)

There are few things related to construction about which there are more misunderstandings and old wives' tales than there are about liquidated damages provisions and their use. The time/cost control system has had liquidated damages provisions as an integral part of its contract documents since the beginning of the development of the system. As a result, provisions similar to those illustrated in Figure 8-3 have been used in a number of states and regions of the United States for private and governmental owners alike. They have been used legally, successfully, and productively over the life of the time/cost control system, which amounts to more than fifteen years. Here are some clarifications of some of the many misunderstandings and old wives' tales:

- There is no state in the United States in which the use of liquidated damages is illegal or unenforceable from a legal standpoint.

- It is not necessary to have liquidated damages provisions in construction contracts accompanied by bonus provisions.
- It has been held by the United States Supreme Court that an owner does not have to prove the amount of his damages as long as both parties agreed to the amount in advance and as long as there is no evidence that the amount of the damages was set with fraud intended by the owner.
- Liquidated damages provisions can be and have been enforced.

The presence of a substantial liquidated damages provision generally will not, of itself, increase the cost of a project. (An important point here, however, is that there should be competitive pressures present to avoid an open invitation to add sums in the construction proposal for the possibility of liquidated damages.) As has been pointed out, many typical construction times can be cut by around 50 percent without any increase in cost or decrease in quality. However, once a schedule and construction contract time becomes further shortened to the point that there will necessarily be real added costs to the contractor, the price of the construction contract itself will increase by those sums, assuming there are strong controls for the owner in the contract for time of completion, such as substantial liquidated damages clauses.

Sometimes attorneys who are not experienced in the practical side of construction management will advise against the use of liquidated damages on the grounds that they are difficult to enforce. It is true that enforcement is difficult at best. This is particularly true of a provision that is not carefully written around the specifics of a project and is accompanied by typical contract time-extension provisions and poor contract administration. But, the important thing that owners and construction managers need to remember is that the purpose of the liquidated damages is not to win some court battle after the fact but to extract the contractor's best efforts to finish the project on time. If a contractor sees that a contract containing liquidated damages provisions is well written, if the amounts of the damages set are realistic as well as significant, if he sees that the owner and architect/engineer or construction manager are serious about the provisions and schedule from the outset, and if he sees that the contract is carefully and fairly administered with good documentation, then he is going to do everything he can to avoid the long court battle or arbitration that might take place while the owner holds the additional retainage that represents the amount of the assessed damages.

The term "liquidated damages" means simply that, the liquidation of damages caused one party by another with payment in appropriate or agreed on amount.

Setting the amount of liquidated damages is, of course, very important. $100 per day damages for a $3 million construction project would be meaningless.

There is a story that persists among many owners and even knowledgeable architects, engineers, and construction managers which says that finishing the project on time is just as important to the contractor as it is to the owner. The story goes, then, that this fact does much more than any form of contract time provisions for getting the work finished within the schedule. This would seem reasonable, and it is basically true that it should be in the contractor's interest to finish the work quickly. However, the stated conclusion does not necessarily follow for several reasons. First, it is unfortunately true that many construction companies, which as a group have exhibited very high mortality rates through the years, are poorly managed when compared to other segments of industry. In turn, some of their actions do not necessarily reflect their own best interests. But a bigger factor in the time-of-completion situation is probably this: As problems of on-time deliveries and coordination develop on a project, usually among subcontractors, a point is frequently reached beyond which efforts to expedite and return to schedule bring a diminishing return to the contractor. The subcontractors may not be receiving critical equipment on schedule; certain trades must complete their work in a given area before other trades can conveniently proceed with their work, and while extra effort

FIGURE 8-3

SPECIMEN SECTION ON LIQUIDATED DAMAGES

(Special Conditions)

LIQUIDATED DAMAGES:

Should the Contractor fail to substantially complete work under this Contract and make the building available for beneficial occupancy on or before the date stipulated for Substantial Completion (or such later date as may result from extension of time granted by Owner), he shall pay Owner, as liquidated damages, the sum of $_____ for each consecutive calendar day that terms of the contract remain unfulfilled beyond date allowed by the Contract, which sum is agreed upon as a reasonable and proper measure of damages which Owner will sustain per diem by failure of Contractor to complete work within time as stipulated; it being recognized by Owner and Contractor that the injury to Owner which could result from a failure of Contractor to complete on schedule is uncertain and cannot be computed exactly. In no way shall costs for liquidated damages be construed as a penalty on the Contractor.

For each consecutive calendar day that the work remains incomplete after the date established for Final Completion, the Owner will retain from the compensation otherwise to be paid to the Contractor the sum of $_____. This amount is the minimum measure of damages the Owner will sustain by failure of the Contractor to complete all remedial work, correct deficient work, clean up the project and other miscellaneous tasks as required to complete all work specified. This amount is in addition to the liquidated damages prescribed above.

and more management would benefit the owner, the contractor sees that his return on the extra costs of more coordination and management—along with some real hard costs of overtime, warehouse purchases, and the like—is not commensurate with the investment. It is at this point, which occurs on almost every project, that the owner's and contractor's interests separate insofar as schedule adherence is concerned. While a great part of the time/cost control system is aimed at planning and assisting contractors in avoiding such happenings, the liquidated damages provisions in sufficient amounts are mandatory in most cases to provide a real incentive to carry out the better-than-normal management procedures the system sets in motion for the contractor.

The amount that is set for damages should be based primarily on the costs the owner will incur if the facility is not completed on time. While the owner of an incomplete facility may not always incur extra expenses on a straight-line, daily basis, any other basis of setting liquidated damages should be avoided.

The interest cost of interim financing, or lost net income, whichever is higher, is a good basis for commercial projects. A $10 million hotel project on this basis had projected interim financing costs of about $2,200 per day toward the end of the project—including allowance for lost return on the cash equity of the owner. It seemed that $2,000 per day for each day the contractor was late in providing beneficial occupancy and $500 per day for each day the contractor was late with final completion would be appropriate. The $500 and the $2,000 would be cumulative should both dates pass without completion in either degree.

In a school project, the added costs of double sessions and transporting students would provide a guide, as would rental costs of temporary facilities, furniture and equipment storage, and double moving costs.

Another public project, such as a hospital or airport, might have its liquidated damages set relative to government bond interest rates. Public authorities who might be the owners of such facilities frequently fund their projects with sale of bonds prior to the commencement of construction. The proceeds are then frequently reinvested in short-term government bonds. Bonds of terms that will provide the projected cash-flow requirements to make progress payments are purchased. If the construction time is extended, near the end of the time when most of the bond funds are needed for the construction but the facility is not available, the owner has lost the income from the reinvestment of the bond funds but still does not have the facility. Therefore, for a $25 million bond issue project, with a current rate on government bills of 4 percent, the owner is damaged by at least $2,739 per day. Thus, $2,500 per day for beneficial occupancy and $500 per day for final completion would seem to be minimum but reasonable amounts in this case.

Once the appropriate amount is selected, it should be set out in the specifications as the amount of damages to be paid by the general contractor for the project. If there are to be a number of separate contracts for a project, one might ask what amounts of damages should be set for these separate contracts, each being smaller than the total project.

Since any one phase of the work can usually cause the entire project to be late, it follows that the full amount of damages should be set in each separate contract. This is particularly true if a separate contract is to be transferred into a general contract, because transfer of such a contract with smaller liquidated damages amounts could leave the general contractor in an untenable situation.

At the same time, it is not always practical to set the damages in a separate contract in full amount. For example, in the $10 million hotel project with damages set at $2,500 per day for beneficial occupancy and final completion combined, if there were a separate early foundation contract of $300,000, the $2,500 per day would be unreasonably high in proportion to the size of that contract. As a rule of thumb, in most instances, damages should probably be held to a maximum of one tenth of 1 percent (0.001) of the contract price per day. Even this could be excessive in an individual case and must be carefully considered from all standpoints in each case.

Progress Payment Scheduling (Figure 8-4)

The provision for scheduling progress payments to the contractor—as illustrated in Figure 8-4—is one of the simplest yet most effective of the time-control contract provisions of the time/cost control system. If the contractor understands it and knows that it will be fairly and fully enforced, this provision alone will keep almost any project from being seriously late. However, it should be considered complementary to the liquidated damages and other time-control provisions and not as a replacement for any of the other provisions with this one exception: With the progress payment scheduling provision in a contract, it would be feasible in many instances to put a maximum on the amount of liquidated damages that could be assessed against the contractor. The maximum might vary from the equivalent of 60 to 120 days of liquidated damages. For example, if damages are set at $1,000 per day for beneficial occupancy and $200 per day for final completion, taking 90 days as a guide, the maximum in liquidated damages that would be assessed against the contractor would be $90,000 + $18,000 = $108,000.

Superintendent Requirements (Figure 8-5)

This provision is meant to further reinforce provisions that encourage better management of the project by the contractor, particularly toward the end and at any times during the contract time that the project is seriously behind schedule. If enforced, it will keep the contractor from a very repugnant practice that has become all too familiar to owners, architects, engineers, and construction managers. That is the practice of moving the project superintendent onto a new project that is starting about the time a subject project is finishing up, but before it is fully finished.

FIGURE 8-4

SPECIMEN SECTION ON PROGRESS PAYMENTS

(Special Conditions)

On or about the first day of each month, and except as may otherwise be required, the Contractor shall submit to Architect an itemized Application for Payment, supported by such data substantiating the Contractor's right to payment as the Owner or Architect may require.

Progress payments will be made for work completed and materials delivered and properly stored in accordance with the General Conditions through the last date listed below. No progress payments will be made after the last date listed below until the final payment except that if extensions in the Contract Time total 30 days or more there shall be additional progress payments for each full 30 days of Contract time extensions.

Under the Base Bid:

July 1, 1972
August 1, 1972
September 1, 1972
October 1, 1972
November 1, 1972
December 1, 1972
January 1, 1973
February 1, 1973
March 1, 1973

Last Date

April 1, 1973

FIGURE 8-5

SPECIMEN SECTION ON SUPERINTENDENT REQUIREMENTS

(Supplementary General Conditions)

Add the following paragraphs and subparagraphs to paragraph 4.9:

The Superintendent shall remain on the project not less than 8 hours per day, five days a week unless the job is closed down due to a general strike or conditions beyond the control of the Contractor or until Termination of the Contract in accordance with these Specifications. The Superintendent shall not be employed on any other project during the course of this Contract.

In the event that any of the following conditions shall exist, the Contractor shall require that his Superintendent be at the Job Site not less than ten (10) hours per day, six (6) days per week:

Should Substantial Completion not be accomplished on schedule.

Should Final Completion not be accomplished on schedule.

Should the CPM network show the Contractor to be 14 or more days behind schedule at any time during construction up until 30 days prior to schedule Substantial Completion.

Should the CPM network show the Contractor to be 7 or more days behind schedule at any time during the last 30 days prior to scheduled Substantial Completion.

In the event that the Superintendent is absent from the job site during the times specified above the Contractor shall pay the Owner, as liquidated damages, the sum of $_____ for each hour of the Superintendent's absence. However, if the Contractor attains both Beneficial Occupancy and Final Completion by the specific dates listed these liquidated damages will be restored to the Contractor.

Owner's Right to Occupy an Incomplete Project (Figure 8-6)

The AIA general conditions do a reasonably good job of spelling out the owner's rights in this area; however, the provision illustrated in Figure 8-6 is a very important detail that should not be overlooked. The conditions of this provision also constitute one further encouragement to the contractor to stay on schedule.

FIGURE 8-6

SPECIMEN SECTION ON OWNER'S RIGHT TO OCCUPY INCOMPLETE PROJECT

(Special Conditions)

Should the Project, or any portion thereof, be incomplete for Beneficial Occupancy or Final Completion at the scheduled date or dates, the Owner shall have the right to occupy any portion of the Project. In such an event, the Contractor shall not be entitled to any extra compensation on account of said occupancy by the Owner or by the Owner's normal full use of the Project. Further, in such an event, the Contractor shall not be entitled to any extra compensation on account of the Owner's occupancy and use of the Project, nor shall the Contractor be relieved of any responsibilities of the Contract including the required times of completion. Such occupancy by the Owner would not, in itself, constitute Beneficial Occupancy or Final Completion.

If the Owner exercises his rights under the foregoing and occupies the full project, then there shall be no liquidated damages on account of failure on the Contractor's part to provide Beneficial Occupancy from that date forward. This provision does not effect, however, any liquidated damages that would be assessed for any period of time between the scheduled date of Beneficial Occupancy and the Date of any such occupancy. Further, this provision would have no effect on liquidated damages assessed on account of late Final Completion.

Extension of Time in Change Orders (Figure 8-7)

This is another supplement to the AIA general conditions that is an integral part of the overall system contained herein for better control on extensions in the contract time.

FIGURE 8-7

SPECIMEN SECTION ON EXTENSION OF TIME IN CHANGE ORDERS

(Supplementary General Conditions)

The Contractor agrees that extensions of time will not be granted for change orders that, in the opinion of the CPM Monitor, do not effect the critical path of the project.

Time for Completion and Specific Dates (Figure 8-8)

Every set of specifications should contain, in the special conditions, the calendar date for substantial completion (beneficial occupancy) and final completion as well as the notice-to-proceed date, on which the completion dates are predicated. Specific dates can refer to several different situations. One would be an interface date with an earlier-awarded or separate contract. In setting such interface dates, it is advisable for the architect or construction manager to allow a little leeway. For example, if an earlier-awarded structural frame contract calls for erection to be completed on June 30, 1972, the general construction contract might list this as July 7, while the actual date is required in the structural contract. Another situation would be an interim occupancy date. For example, a manufacturer might require early occupancy of a space

FIGURE 8-8

SPECIMEN SECTION ON TIME FOR COMPLETION AND SPECIFIC DATES

(Special Conditions)

TIME FOR COMPLETION:

Each Bidder's attention is directed to the fact that the building is urgently needed by the Owner and that time is of the essence; for this reason, it shall be agreed that the Contractor shall complete all work under the Contract within the time established under paragraph titled Specific Dates below. Should the award of the contract and notice to proceed be delayed beyond the date set out below, all subsequent dates shall be set one day later for each day that the award of Contract or Notice to Proceed shall be delayed; whichever shall come first.

SPECIFIC DATES:

The schedule below contains certain specific dates in addition to date of Notice to Proceed and Time for Completion. These dates shall be adhered to and are the last acceptable dates unless modified by mutual agreement between the Contractor and the Owner. All dates indicate midnight unless otherwise stipulated. The only exceptions to this schedule are defined in the Supplementary General Conditions under EXTENSIONS OF TIME.

a. Award of Contract or Notice to Proceed	June 2, 1972
b. Foundations ready for Structural Frame Erection to begin	June 23, 1972
c. Structural Frame Substantial Completion	July 7, 1972
d. Substantial Completion (Beneficial Occupancy)	September 15, 1972
e. Final Completion (Except for Warranties)	October 15, 1972

for equipment installation. This date should be called out in the specifications and separate liquidated damages can be specified for the special occupancy.

Once the design and construction sequencing has been determined and the contract executed between the owner and the contractor or contractors, the major task for the architect/engineer or construction manager becomes contract administration. Contract administration is covered in parts of Chapters 10 and 12; however, it is important for the reader to relate certain aspects of contract administration to this subject—time control.

All too often the owner does not get what he has contracted for. This is particularly true of time provisions, because resident representatives, owner's representatives, or architect/engineers do not carefully follow the owner's rights in the contract, inadvertently let contractors out of responsibilities, fail to impress the contractor with the fact that the contract will be enforced, or fail to understand the philosophy of the contract in the first place.

With respect to general contracts in particular, the role of the contractor has changed over the last four or five decades, particularly since World War II. One often hears the criticism of a general contractor that "he is just a broker," meaning that he subcontracts everything. This is not a reasonable criticism; in fact, playing that role and being surety and coordinator is the proper role for a general contractor. Anyone who thinks that a general contractor carries a large force of mechanics, laborers, and trade specialists will almost invariably be fooled. But the fact that most contractors do not carry such a force is by no means detrimental to an owner. In fact, it may be in an owner's best interest, since it leaves the main tasks of coordination and management to the contractor and the various trade work to specialists in each field.

The philosophy of a modern, single-responsibility general contract should be that, because of the contractor's construction expertise—which owner and the architect/engineer do not have—he accepts the risk of completion within an agreed on amount and time. If the general contractor does not accept this risk, meaning that this risk must be properly placed on him by the contract, then the general contractor has no significant role.

It is, then, within this philosophy or concept that rulings on the contractor's requests for contract time extensions should be made by the architect/engineer or the construction manager.

Basically, the contractor can and should be expected, in placing his proposal for a specific time of completion, to take into account those things that his specific expertise would tell him he should allow for.

No one in the industry or related professions should be in a better position than the experienced contractor to know that allowances must be made for ordering and fabricating, normally inclement weather, coordination, labor problems, etc.

Therefore, the architect/engineer or construction manager will be derelict in his duty to the owner if he allows extensions in the contract time for

- Normally inclement weather or weather claims based on only a segment of the construction calendar time period
- A labor strike that does not affect the ongoing critical path work
- Normal and reasonable reaction time by the architect/engineer, construction manager, and owner
- Failures on the part of his subcontractors, whether he selected them originally or accepted their contracts under a full transfer
- Failures in delivery or problems with product and equipment manufacturers

When a contractor submits a request for contract time extension, the architect/engineer or construction manager should do the following:

1. As a minimum, immediately simply acknowledge receipt of the request.
2. If it is clear that the request will be rejected, reject it immediately, in writing.
3. If it is clear that the request will be granted in full or in part, and in the latter case, if the number of days is then known, grant the request or approved part of the request immediately, in writing. (Note that in most situations, the architect/engineer or construction manager will be required to obtain the owner's concurrence for granting extensions.)
4. If the request has to do with weather, unless it is the result of actual damage to the work caused by windstorm, lightning, or the like, state that the request will be ruled on at the end of the job. If, however, it is clear then that the request will be denied, the action should be as in 2 above. (Sometimes it may be advisable to make "tentative interim" rulings on weather so that realistic updates of the critical path method schedule can be made and to avoid undue progress payment hardship on the contractor. But the architect/engineer and construction manager should remember that it will be most unusual for any calendar period of six months or longer to actually have net inclement days beyond the norm; such conditions rarely occur for more than a very few days.)
5. If the request cannot be ruled on immediately because of inadequate information or because insufficient time has passed to observe effects, state that to the contractor, and thereafter make the ruling, as soon as possible, in writing.

CHAPTER 9

Productive Use of the Critical Path Method and Other Computer-Assisted Scheduling Techniques

The critical path method of scheduling, which will be referred to throughout this chapter as CPM, and other computer-assisted scheduling techniques such as PERT have many good applications both in the construction of buildings and in design and programming of construction projects. Likewise, these techniques may be used in other related activities such as planning, systems development, etc.

However, the role of CPM in the time/cost control system is that of scheduling the construction itself and related offsite fabrication. At the same time, the use of CPM principles are highly recommended for the planning and program management processes also.

While having its beginning well before 1960, CPM began to be a fairly popular method of construction scheduling in the early 1960s. As in the case of so many other more sophisticated techniques, however, some individuals who have made one or two attempts at using the system have often made the mistake of placing total reliance on CPM, to the exclusion of other practical time-control techniques such as the contract provisions discussed in Chapter 8. These users later concluded that CPM was of no particular value or at least not worth the cost of implementing the system. There have been many other misuses of CPM, but it has made positive contributions in too many projects for it not to be considered a practical and worthwhile time-control technique.

A major point that should be kept in mind about CPM is that it is one of the tools of the construction manager or architect/engineer for time control in construction; it is not the only tool to be utilized, nor is it likely to be successful unless other tools and techniques are combined with it, including techniques that would give the contractor involved the incentive to utilize the CPM program.

The time/cost control system has several key principles regarding the use of CPM. These are as follows:

1. CPM must be put into use in such a way as to have bidding contractors utilize the CPM system and the services of the appointed CPM consultant prior to placing their proposal, at no expense to them other than the cost of their own time.

2. CPM should be set up as a tool for the contractor, but one that he is not forced to utilize in detail, nor should it be allowed to supersede the contractor's contractual completion date or other requirements.

3. The CPM system should be maintained throughout the project, whether or not the contractor chooses to use it. It will be needed as a reporting system to the owner, construction manager, and architect/engineer, and it will be useful as a basis for making contract time-extension rulings by the construction manager or architect/engineer. (Also for rulings by the CPM consultant; see provisions for contract time-extension rulings in Chapter 8.)

Figures 9-1 shows a specimen specification section for scheduling and reporting covering the use of CPM as it is utilized in the time/cost control system for scheduling the construction work of the individual contractor.

Here is how the system is implemented and how the specimen specification referred to above is administered:

There should be one or more individuals in the construction management or architectural/engineering firm sufficiently experienced in construction operations sequencing to develop the logic for the project schedule in substantial detail. Prior to the completion of the contract documents, preferably as early as the design development phase of the architect/engineer's work, this construction scheduling specialist should commence work with the CPM consultant.

The CPM consultant, as can be seen in Figure 9-1, is appointed by the construction manager or architect/engineer in the specifications and is cast in a role similar to that in which a testing laboratory has been traditionally cast.

During the predesign project analysis, the project manager of the construction management firm or architect/engineer, along with key architectural/engineering design personnel, will have developed the basic fundamentals of the project schedule as a part of the construction management plan. At that time, any major constraints that would cause early awards of constraint activity contracts will have been identified through the "think tank" process, and the same process will have determined the basic time span that is feasible for the contract time.

The first activity of the construction scheduling specialist and the CPM consultant will be to reevaluate the schedule of the management plan. This reevaluation, if done during the design development phase, will allow sufficient time in most projects for corrections in the schedule that may be indicated by the detailed CPM analysis.

FIGURE 9-1

SCHEDULES AND REPORTS

Division 1 / Section 1E

GENERAL REQUIREMENTS:

1. The work under this Contract will be scheduled and reported by the Critical Path Method, using activity on arrows, precedence diagraming, or activity-on-node. The following publication is cited as a reference for CPM work under this contract: "Project Management with CPM and PERT," J. J. Moder and C. R. Phillips, Reinhold Publishing Corporation, New York, 1970.

2. The Architect has appointed ______________________ as the CPM Consultant for this project. The Owner will pay the CPM Consultant for its services.

3. The services of the CPM Consultant and the schedules prepared by the CPM Consultant are made available to the Contractor as an aid. It is intended that these schedules will reflect the Contractor's actual construction plan. However, the existence of schedules, networks, vector charts, and services of the CPM Consultant do not in any way relieve the Contractor of the responsibility of completing the work within the scheduled time for both Beneficial Occupancy (Substantial Completion) and Final Completion, nor do they in any way relieve the Contractor of the responsibility of meeting any of the other requirments of the Contract Documents.

PREBID SERVICES AND REQUIREMENTS:

4. The CPM Consultant will, prior to the receipt of bids:

(a) Prepare a provisional preliminary network, which is not contained in this specification.

(b) Take part in a prebid conference, familiarizing prospective bidders with the Critical Path Method and the provisional preliminary network. The prebid conference is scheduled for 2:00 PM, EST, June 26, 1973, at the office of the Architect.

(c) Be available to each Bidder for this Contract for up to four hours of individual consultation, to assist the Bidder in the preparation of a revision of the preliminary network as specified. The CPM Consultant shall make himself available to the respective Bidders at a mutually agreeable time in the City of Atlanta, Ga. The respective bidder should make an appointment with the CPM Consultant immediately following the prebid conference and must allow the CPM Consultant a reasonable and adequate time to make required revisions. If a bidder desires further services, he shall make suitable arrangements for such services with the CPM Consultant at no additional cost to the Owner or Architect.

5. Each Bidder shall submit with his proposal either the provisional preliminary CPM network as originally prepared by the CPM Consultant or as revised in accordance with each bidder's own plan of construction, and will, in so doing, accept the network submitted as one workable schedule for the project. If revised by a prospective bidder, the end dates, i.e., dates of Substantial Completion (Beneficial Occupancy) and Final Completion and any specific dates listed in paragraph 6 of the Special Conditions may not be changed. The preliminary CPM network submitted with the proposal shall then be referred to in the Proposal Form and thereafter as the Preliminary CPM Network.

PRIMARY SCHEDULE:

6. The Contractor and the CPM Consultant shall do the following after receipt of Notice to Proceed:

(a) The CPM Consultant shall conduct a one-day training session on the use of CPM for this project. The Contractor will see that his own managers and superintendents and those of his Subcontractors attend this session.

(b) Within ten (10) calendar days after issuance of Notice to Proceed or award of the Contract by the Owner, whichever comes first, the Contractor and the CPM Consultant shall meet together to expand and immediately complete the network. The Contractor and major Subcontractors shall provide the CPM Consultant with such additional information as is required to prepare a complete detailed network. The level of detail of the network shall be such that no activity shall be longer than 25 days, except for procurement activities. The Contractor and his Subcontractors shall provide full information and assistance to the CPM Consultant according to a predetermined schedule.

(c) This additional information shall include a dollar value (cost) for each work activity. The cost shall include labor, material, and pro rata contribution to overhead and profit. The sum of all activity costs shall be equal to the total contract price. Each activity cost shall be coded with a cost code corresponding to the sections of these specifications so that subtotals for each division of the work can be prepared.

7. Within twenty-four (24) days from the Notice to Proceed, the CPM Consultant and the Contractor shall provide the network and computer printout to the Architect for review and approval. Within thirty (30) days from Notice to Proceed, the CPM Consultant shall provide the required copies of the approved network and computer printout. This schedule shall be referred to as Detailed Network Number 1.

NETWORK CONTENT:

8. The network must contain detailed representation of all significant aspects of the construction plan, including but not restricted to site preparation, structural work, interior and exterior finishes, electrical and mechanical work, and acquisition and installation of special equipment and materials. For all equipment and materials fabricated or supplied especially for this project, the network shall show a sequence of activities including preparation of shop drawings, approval of drawings, shop fabrication, and delivery. A time-phased graphic representation demonstrating duration and overlap of significant activities must be included. At the CPM Consultant's option, this graphic representation may be included as part of the CPM printout, rather than the network.

9. The initial and subsequent CPM computer printouts shall include the following minimum information for each activity: Activity Number (or I, J); Activity Description; Estimated Duration in days; Early and Late Start dates, Early and Late Finish dates; and cost. The initial schedule shall indicate an early completion date for the project that is no later than the project's required completion date. All activity dates shall be given in calendar days.

10. The CPM schedule shall consist of a list of the project activities sorted and tabulated in the following ways:

(a) By Predecessor Event Number as a major sort and by Successor Event Number as a minor sort, or by Activity Number
(b) By Total Slack, from the least to the most
(c) By Late Start date, in chronological order
(d) By Activity Number, including cost value and value in-place, to be used as itemized application for payment

Actual start and finish dates should be indicated for each activity that has started or finished. Dummies and finished activities shall be omitted from Total Slack and Late Start sorts.

UPDATES:

11. On the first Wednesday of every month, the Contractor's superintendent and the Architect's representative shall meet at the jobsite for the purpose of obtaining from the Contractor (following his subsequent meeting with all concerned subcon-

tractors and suppliers) up-to-date and accurate progress input data. This data shall have the concurrence of the Architect's representative and shall be promptly forwarded to the CPM Consultant. The CPM Consultant, immediately upon receipt of this data, shall make a network revision and computer update, with distribution as follows, within five days:

Contractor	6
Architect	2
Owner	1

12. The CPM Consultant shall include with each revision and update an analysis of the progress of the project as reflected in the current update. The CPM Consultant shall, promptly upon the availability of this update and analysis, schedule a meeting at the jobsite with the Contractor's superintendent and Architect's representative to review critical features of this update.

PROGRESS PAYMENTS:

13. The updated CPM schedule shall be used as the itemized application for payment referred to in the Special Conditions section of these specifications.

REVISIONS:

14. In the event the Contractor makes any change in his plan of operation, he shall provide the CPM Consultant with the details of this revised plan so that the changes can be incorporated into the CPM schedule. If these changes require extensive revision of the logic and time deviations, the Contractor shall pay the CPM Consultant for the cost of incorporating the changes.

SPECIFIC DATES:

15. Adhere to the specific dates as required by the Special Conditions section of these specifications.

The construction scheduling specialist and CPM consultant, working together, will develop a diagrammatic CPM vector network similar to that illustrated in Figure 9-2. For most projects, this will consist of from 50 to 250 major activities.

During the remaining design development and construction document preparation phases of the architect/engineer's design work, the construction manager and the architect/engineer can make use of the rough CPM network diagram for reference as needed on many points. By the time the construction contract document preparation phase is drawing to a conclusion, the scheduling specialist and the CPM consultant should have refined the CPM network into the provisional preliminary network, and this network should be reproduced and bound into the specifications as an exhibit in the scheduling and reporting section of the specifications.

In turn, the provisional preliminary network will be one of the major subjects of discussion at the prebid conference, at which the CPM consultant should first brief all bidders on the general principles of CPM and its use.

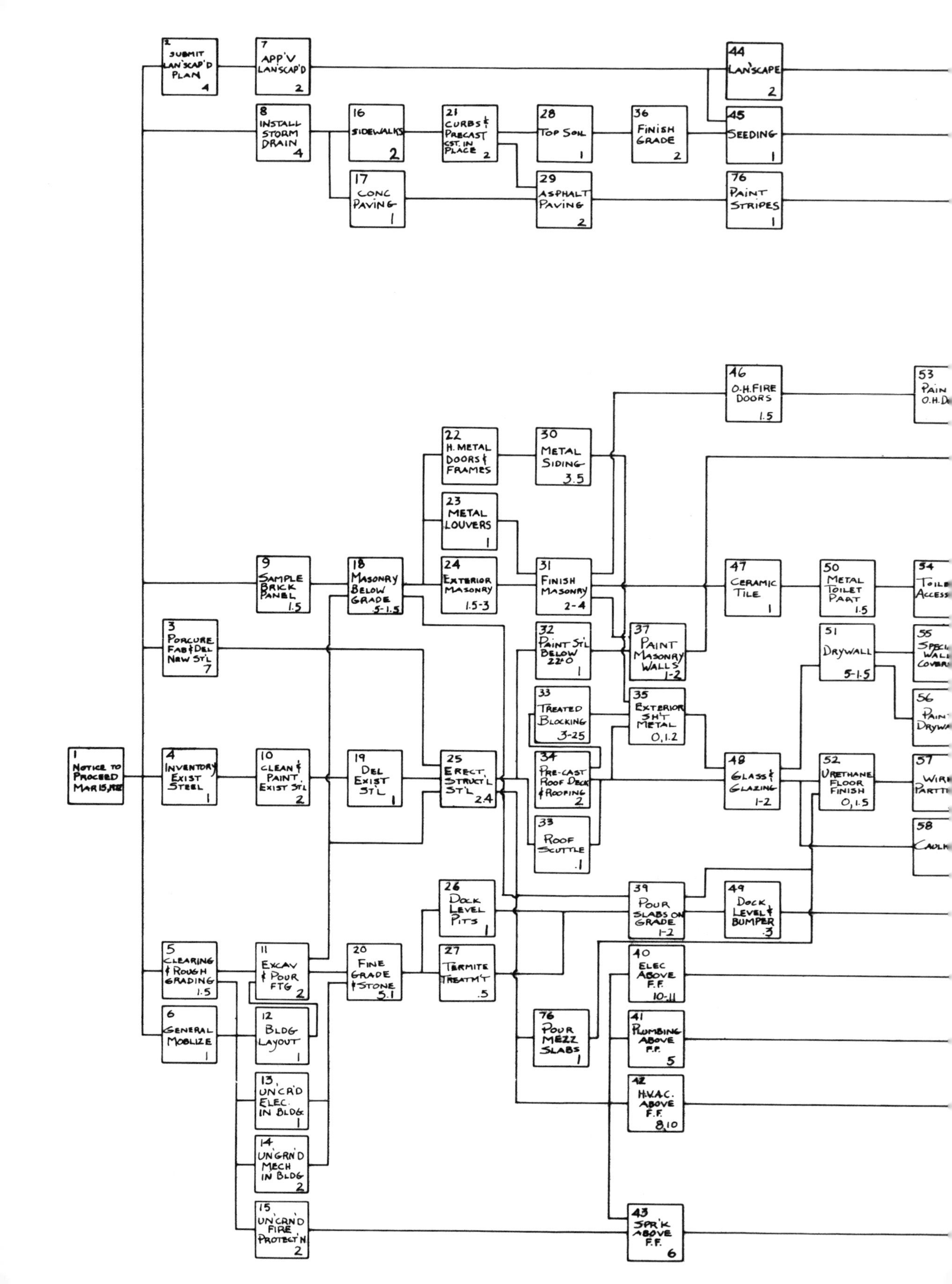

1 NOTICE TO PROCEED MAR 15, 1982
2 SUBMIT LAN'SCAP'D PLAN 4
7 APP'V LAN'SCAP'D 2
44 LAN'SCAPE 2
8 INSTALL STORM DRAIN 4
16 SIDEWALKS 2
21 CURBS & PRECAST CST. IN PLACE 2
28 TOP SOIL 1
36 FINISH GRADE 2
45 SEEDING 1
17 CONC PAVING 1
29 ASPHALT PAVING 2
76 PAINT STRIPES 1
46 O.H. FIRE DOORS 1.5
53 PAIN O.H. D
22 H. METAL DOORS & FRAMES
30 METAL SIDING 3.5
23 METAL LOUVERS 1
9 SAMPLE BRICK PANEL 1.5
18 MASONRY BELOW GRADE 5-1.5
24 EXTERIOR MASONRY 1.5-3
31 FINISH MASONRY 2-4
47 CERAMIC TILE 1
50 METAL TOILET PART 1.5
54 TOILE ACCESS
3 PORCURE FAB & DEL NEW ST'L 7
32 PAINT ST'L BELOW 22'-0 1
37 PAINT MASONRY WALLS 1-2
51 DRYWALL 5-1.5
55 SPECI WALL COVERI
33 TREATED BLOCKING 3-25
35 EXTERIOR SH'T METAL 0,1.2
56 PAIN DRYWA
4 INVENTORY EXIST STEEL 1
10 CLEAN & PAINT EXIST ST'L 2
19 DEL EXIST ST'L 1
25 ERECT STRUCT'L ST'L 2.4
34 PRE-CAST ROOF DECK & ROOFING 2
48 GLASS & GLAZING 1-2
52 URETHANE FLOOR FINISH 0,1.5
57 WIR PARTIT
33 ROOF SCUTTLE .1
58 CAULK
26 DOCK LEVEL PITS 1
39 POUR SLABS ON GRADE 1-2
49 DOCK LEVEL & BUMPER .3
5 CLEARING & ROUGH GRADING 1.5
11 EXCAV & POUR FTG 2
20 FINE GRADE & STONE 5.1
27 TERMITE TREATM'T .5
40 ELEC ABOVE F.F. 10-11
6 GENERAL MOBLIZE 1
12 BLDG LAYOUT 1
76 POUR MEZZ SLABS 1
41 PLUMBING ABOVE F.F. 5
13 UN'CR'D ELEC. IN BLDG 1
42 H.V.A.C. ABOVE F.F. 8,10
14 UN'GRN'D MECH IN BLDG 2
15 UN'CRN'D FIRE PROTECT'N 2
43 SPR'K ABOVE F.F. 6

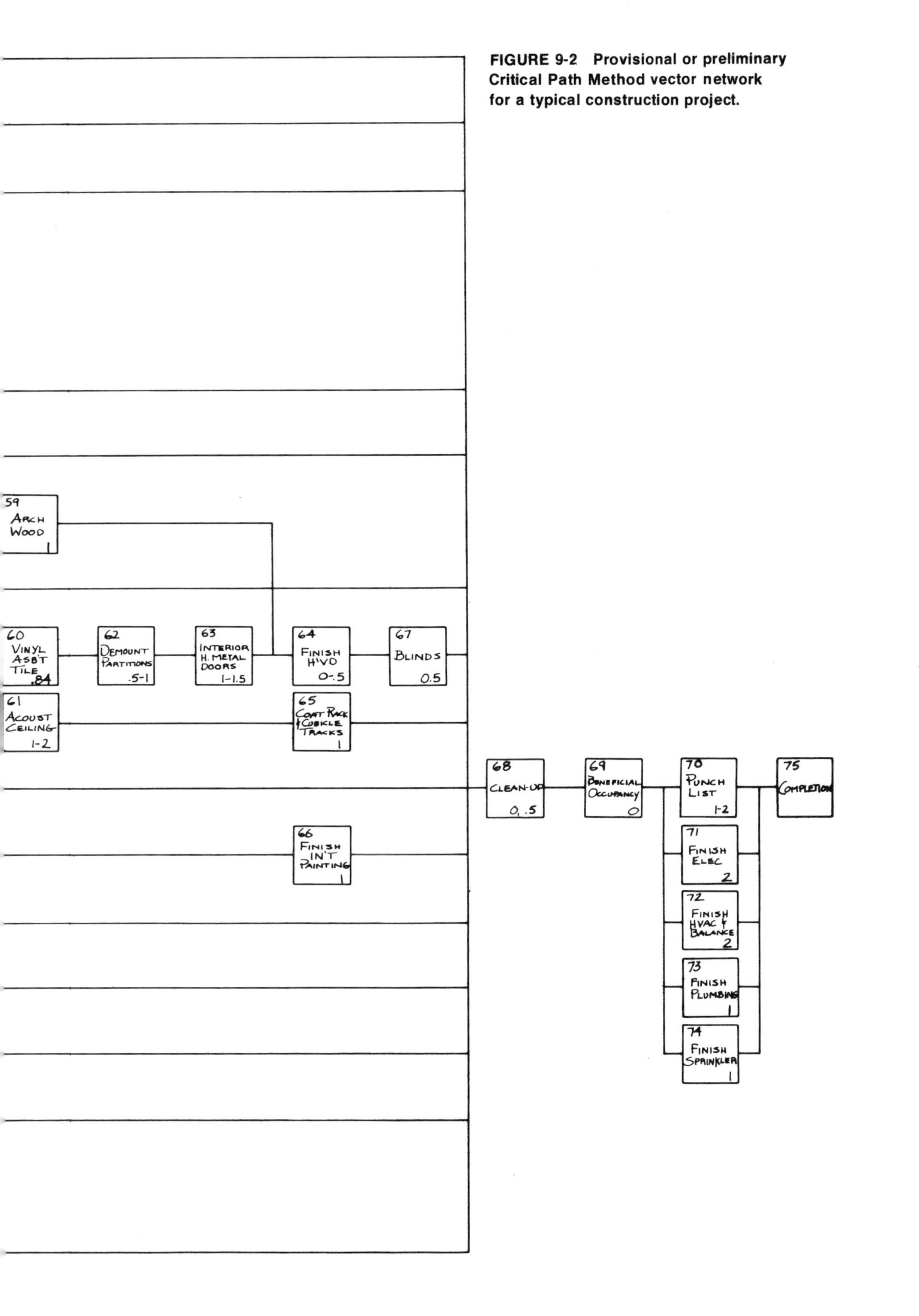

FIGURE 9-2 Provisional or preliminary Critical Path Method vector network for a typical construction project.

Further, also in the prebid conference, the provisional preliminary network and its logic should be discussed in some detail with the attending bidders. If a major flaw in the logic of the schedule is revealed in this discussion with contractors, a revised provisional preliminary network should immediately be prepared and issued in an addendum.

Assuming that no major flaws in the logic are turned up in the prebid conference, the respective bidders at that conference should be directed to make individual appointments with the CPM consultant. Each bidder can then subsequently meet with the consultant (whose time should be provided at the cost of the owner, constituting no outside cost to the individual bidder) in order to review the schedule in more detail for the respective bidder.

Subsequently, each respective bidder must determine that he will accept, as one workable basic schedule, the provisional preliminary network or a preliminary network that he and the CPM consultant work out for him. The specimen specification on scheduling and reporting stipulates that each bidder may change the provisional preliminary network in any way he chooses as long as he does not change the contract beneficial occupancy and final completion dates or any contract interface dates (a date for completion of another contract that has been separately awarded, whether or not it is to be transferred into the general contract under discussion).

Therefore, upon the placement of the bid, each bidder has either accepted the provisional preliminary network as a workable schedule, thereby making it the "preliminary network", or he will have developed his own preliminary network which he will submit with the bid.

If values of activities have been assigned to the CPM schedule activities, then this preliminary network will replace the traditional cost breakdown submitted by contractor after the award and will, once it has been approved by the construction manager or architect/engineer, form the basis for progress payments. In any case, the accepted preliminary network will provide the basis for future evaluation by the construction manager or architect/engineer, as well as the CPM consultant, in making any necessary rulings on contract time extensions.

Once the contract has been awarded, the CPM network will be expanded in the form of a computer printout in accordance with the specimen specifications on scheduling and reporting. As can be seen in the specimen specification, a specific day of the month is stipulated in the specification (assuming monthly updates; however, updates may be biweekly, monthly, or at other intervals). A further stipulation is that, by a certain period of time after the award of the contract, the CPM consultant, having consulted with the general contractor, will expand the number of activities from that which was used in the preliminary and provisional preliminary scheduling to at least the minimum number called for in the specifications.

Referring to the case histories in Chapter 1, the Atlanta Stadium project (a highly accelerated, major construction project), had about 200 activities in the provisional and preliminary network. This was expanded to 3,000 activities after the award, and the updates were made every fourteen calendar days.

In the Delta Air Lines Computer Center, also a case history from Chapter 1, the preliminary network scheduled 50 activities. This was expanded to 500 activities, with monthly updates, after the award.

See the specimen specification in Figure 9-1 for the requirements on all parties for periodic updates during the construction phase. On a specified standard date after each update, usually within five calendar days after the required jobsite meeting or after the gathering of input data, the update should be turned over to the contractor in a meeting at which the CPM consultant will brief all concerned on the status and problems.

An illustration of how the time/cost control application of CPM scheduling techniques can pay big dividends in time savings is contained in the Atlanta Stadium project. This heavy structure was to be located on piles, and there was a significant quantity of earth moving that had to take place in the area of the grandstands and the playing field. Most experienced construction superintendents would have scheduled the grading to be completed and then the pile-driving operation to immediately follow. However, the CPM analysis indicated that if this sequence were followed, it would be impossible to complete the project within the required twelve-month period. In turn, it was indicated that the pile-driving operation had to take place on a three-shift basis concurrently with the grading, with some pile driving through overfill and some piles driven from a base point of lower than the pile cap elevation, with fill being placed around individual piles later. In retrospect, it was easily seen that this one point, uncovered in the prebid CPM scheduling, kept about thirty to forty-five days from being lost at the outset of the project. Not only was the prebid CPM schedule responsible for uncovering this fairly simple fact, but the way in which it was used in the time/cost control system, namely before the bid was placed, meant that this information was known in time, before it was too late to recover the time.

A good example of how the CPM preliminary network can be used in contract time-extension rulings is represented by another stadium project, Sanford Stadium for the University of Georgia. The Sanford Stadium project was a rebuilding on an existing stadium site; the old stadium had to continue in use until the day before the construction work commenced. Consequently, it was not feasible to take detailed subsurface tests in the area of the old existing grandstands. The tests were taken around the stadium area, but—as it turned out—they did not reveal a very unusual and complicated condition existing under the grandstands. Apparently, in the construction of the old grandstands (in fact, by the memory of the author's grandfather), there had been extensive blasting of rock. This blasting, which apparently was hardly done in a professional manner, not only dislocated a series of boulders but caused a series of vertical and near vertical faults in the rocks and boulders underlying the old grandstand. This condition was further complicated by the fact that, accompanying the hard blue granite boulders, were other decomposing rock formations. These were too dense for the proposed caisson augers to penetrate, but they did not provide a sufficient spread footing base for the heavy loads of the grandstand columns. These two unusual conditions, which occurred to a major extent only on the south side of the playing field, caused a loss of some thirteen weeks from the original schedule in the subsurface work for the south stands.

The superstructure of the project called for cast-in-place concrete bents, with precast risers for the seating. There were twenty-five very large bents on each side of the field, and the general contractor had scheduled in his prebid schedule (and had in turn ordered) four engineered steel forms, two to be used on each side of the field. When the condition of the subsurface of the south side became known, the architect/engineer/construction manager directed the contractor to move the south side forms to the north side and to make an optimum utilization of all four sets there. By the time the form work on the north side had been done, the foundation problem on the south side had been solved and all four sets were moved back to the south side. Subsequently, the contractor requested a thirteen-week extension in contract time, because it had taken thirteen weeks longer than scheduled to complete the subsurface work on the south side. Without the CPM schedule, any construction manager or architect/engineer in a similar situation would have been virtually forced into approving the request. However, the CPM network, as borne out by the CPM consultant, showed that the critical path for the project went through the foundation work for the first twenty-eight days of the project, and then the critical path became the utilization of the steel forms in the casting of the concrete bents.

By recording the use of the steel forms for the concrete bents, all four being first used on the north side and then all four being transferred for use on the south side, it was shown, and confirmed by the CPM consultant, that only thirty-one days were lost. In turn, a ruling was made to that effect, and the contractor completed the project within thirty-one days of the original date of beneficial occupancy.

CPM scheduling is, of course, somewhat complicated when there are separate contracts on a project. However, this does not change the principles involved in CPM nor in its method of use in the time/cost control system.

When a number of separate contracts are employed, the overall project schedule is first of all developed for separately awarded specialty work as well as the general construction work. From this, interface dates are developed. Here it should be noted that interface dates for two contracts with an interdependent activity should not be the same insofar as contractual requirements are concerned. For example, if an early awarded structural frame contract is to be completed by the date of March 1, then in the contract time requirements for the structural frame contract, that date (March 1) should be required for the completion date for the erection of the frame. However, for the general construction contract, whether or not the structural frame contract is to be transferred into it subsequently, a date containing some leeway should be stipulated, such as March 10 or March 15, depending on the size and complexity of the project.

The preceding is simply a practical legal move and is not necessarily part of CPM technique.

Once the overall project schedule has been developed, an individual network for each of the different contracts should be developed. Afterwards, the same procedure should be followed for each contract, as has been described above in the use of CPM and the specimen specifications on scheduling and reporting.

It is recommended that construction managers and architect/engineers, as well as students interested in the subject, read James J. O'Brien's *Scheduling Handbook*, copyrighted in 1969 and published by McGraw-Hill Book Company, and his *CPM in Construction Management*, 2nd ed., copyrighted in 1972 and also published by McGraw-Hill.

Other good references on CPM are *Project Management with CPM and PERT* and *Management Program Planning and Control with PERT, MOST, and LOB*.

CHAPTER 10

Phased, Separate, and Transferable Contracts

In the late 1960s and early 1970s, the terms "phased construction" and "fast track" became popular jargon in construction management and architectural and engineering circles. Like so many jargon words, the terms have been confused with other terms, and the technique of phasing construction, which has fine potential under the proper conditions, has frequently been proposed or applied in ways that were not in the best interests of the owner.

One point of confusion is that "phased construction" has been identified with the term "construction management." While it certainly requires more management to utilize phased construction than a single general contract, construction management is needed on all construction projects whether or not they are phased, and "phased construction" cannot be considered synonymous with "construction management."

When considering phased, separate, and transferable contracts, it should first be recognized that most owners are best served by a single-responsibility lump-sum general contract. It is true that there are many instances when other methods of managing the construction program will be in the best interest of the owner, but that should be a point of reference for almost any project.

There is a widely held misconception that any project will be accelerated by an early award of site and foundation work, with this phase of the construction starting prior to the completion of design. While this technique has excellent applications in certain projects, one can hardly say that all projects can be successfully accelerated in this manner. As a matter of fact, other activities are far more likely than site and foundation work to constitute a constraint in the typical project. Availability of the site; owner's administrative procedures; fabrication and erection of the structural frame; fabrication and delivery of heavy electric switchgear and major mechanical equipment; delivery and installation of elevators; delivery and installation of escalators, laboratory casework, and other special equipment; and off-site utility and site access work are all far more frequently found to constitute real constraints in a large number of projects. Only in the high-rise project with complicated subsurface conditions is foundation work likely to constitute a constraint. And only on sites that have unusual site conditions or an extraordinary amount of earth moving as a requirement is site work likely to constitute a constraint.

By scheduling the design work for site work and foundations to proceed out of sequence, the architect/engineer's work is disrupted and becomes less efficient. More management effort and cost is required to maintain the owner's competitive position. And a contractor is always at a disadvantage, to some extent, when he undertakes any portion of work with less than complete drawings and specifications and is unable, in turn, to plan his work ahead.

Some of the different conditions that are likely to make phased or separate contracts desirable or necessary (whether or not they are subsequently transferred into a general contract) are as follows:

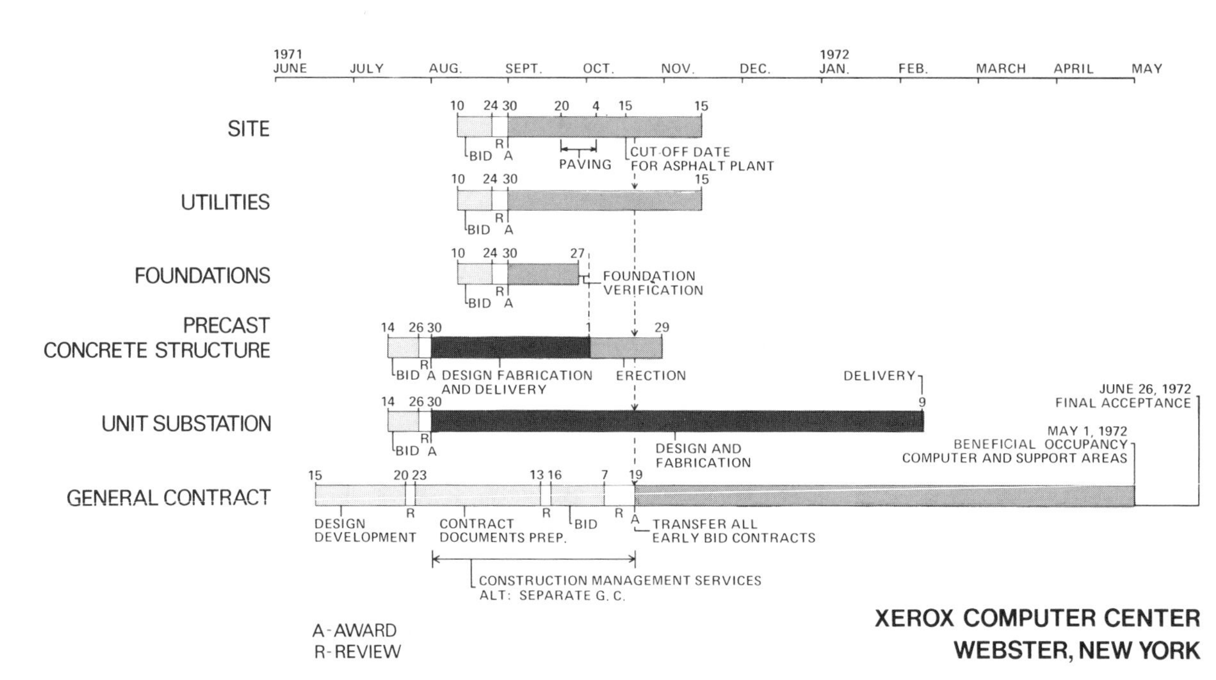
1971 JUNE
JULY
AUG.
SEPT.
OCT.
NOV.
DEC.
1972 JAN.
FEB.
MARCH
APRIL
MAY
SITE
10 24 30 20 4 15 15
BID R A
PAVING
CUT-OFF DATE FOR ASPHALT PLANT
UTILITIES
10 24 30 15
BID R A
FOUNDATIONS
10 24 30 27
BID R A
FOUNDATION VERIFICATION
PRECAST CONCRETE STRUCTURE
14 26 30 1 29
BID R A DESIGN FABRICATION AND DELIVERY
ERECTION
DELIVERY
UNIT SUBSTATION
14 26 30 9
BID R A
DESIGN AND FABRICATION
JUNE 26, 1972 FINAL ACCEPTANCE
MAY 1, 1972 BENEFICIAL OCCUPANCY COMPUTER AND SUPPORT AREAS
GENERAL CONTRACT
15 20 23 13 16 7 19
DESIGN DEVELOPMENT
R
CONTRACT DOCUMENTS PREP.
R
BID
R
A
TRANSFER ALL EARLY BID CONTRACTS
CONSTRUCTION MANAGEMENT SERVICES
ALT: SEPARATE G. C.
A-AWARD
R-REVIEW
XEROX COMPUTER CENTER
WEBSTER, NEW YORK

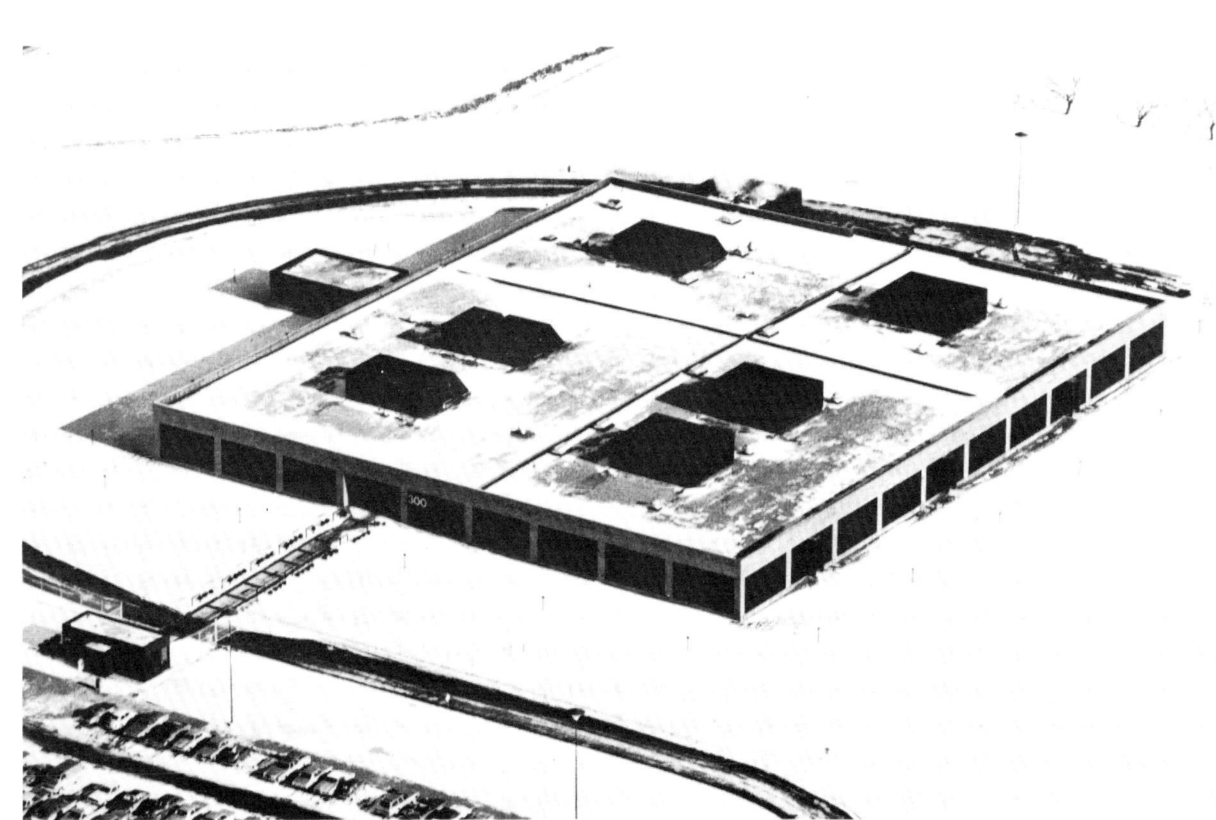

Project schedule illustrating phased construction. Note that several of the early-awarded contracts are subsequently transferred into the general contract once the general contract has been bid and awarded. The illustrated project was a computer center for Xerox Corporation in upstate New York. (Heery & Heery, Architects & Engineers.)

Xerox Corporation, Webster, New York. (Heery & Heery, Architects & Engineers.)

1. In accelerated projects, when one or more constraint activities (as discussed above) are identified, then it will usually be very advantageous to obtain early bids for or to negotiate only for the contracts for these constraint activities (or, in the case of administrative activities, site availability, etc., to deal with them in the appropriate manner in the individual case).

2. In the case of the utilization of industrialized building systems, there is often an advantage to obtaining early bids for this work. However, the construction manager and architect/engineer should use caution in this decision, for it is not uniformly true that early bids are necessary in the utilization of industrialized building systems. Further, there are those circumstances when it is advisable to obtain early bids with a long price hold—i.e., 60 to 120 days or more—and not actually make early or phased awards for all of the systems work. (For example, if the SCSD system is being utilized on a project and the project is to be accelerated, it may be that the fabrication and erection of the structural frame are identified as the only early award activities that are really needed. However, when preparing the contract documents for a systems project, as has been discussed in Chapter 6, it is very advantageous and sometimes necessary to have prior knowledge of who the various systems component manufacturers are to be in order to complete the contract documents in the best and simplest form. Therefore, under these circumstances, early bids might be received for the structural work component, the mechanical component, the ceiling/lighting component, and the interior partitions. Bids would be received on the basis of having rights for the owner to award the contract up to 120 days after the receipt of bids. At the appropriate time, an early award would be made of the structural component contract for later transfer into the general contract. However, contracts for the other three components would not be awarded. Instead, contract documents would be completed based on the successful bidder in each case, and each bidder would merely be notified that he was the apparent low bidder. At the time of the receipt of bids or negotiation of the general contract, the remaining proposals would be transferred into the general contract or would be simultaneously awarded and transferred. At the same time, the previously awarded structural system contract would be transferred to the general contract.)

3. Statutory requirements of certain political subdivisions, notably the state of New York, wherein separate bids and contracts are required on public work for general construction, heating, ventilating and air conditioning systems, plumbing systems, electrical systems, elevators, and sometimes other portions of the construction. These statutes can only be considered special-interest statutes, being promulgated primarily by specialty contractor groups and some labor unions. They are definitely not in the best interests of the government involved or those of its taxpayers. Nonetheless, they do exist and must be dealt with in those situations where they are unavoidable. In the state

of New York, the statutes will not allow the subsequent transfer of separately bid contracts into a general contract. Consequently, these separate contracts must remain separate throughout the project. In the state of Ohio, which requires a minimum of three separate bids—covering general construction, mechanical systems, and electrical systems—for public works, it has been allowable to transfer the mechanical and electrical contracts to the general contracts once they have been bid and awarded separately. Other states and political subdivisions have different variations on these laws. However, a number of states and political subdivisions have wisely kept such laws off their books. The important point is not so much that it is always in a given owner's best interest to have a single general contract but that a governmental or any other owner should not have its hands tied in utilizing the construction management plan that is best for the individual project. There can, in fact, be other reasons for wanting to have separately bid and awarded contracts. These have been discussed above and are further discussed below.

4. Projects that are so large that it is difficult, in the locale of the project, to obtain adequate competition for a single general contract covering all the work.

5. Larger projects having a projected construction period of around 1½ or 2 years or longer and which contain certain items that the general contractor cannot realistically price or order until roughly 6 months or more after the date of award. Because of this condition, the contractor will have to protect himself with generous contingency allowances and escalation allowances. Therefore, owners will probably receive a better price by separating these items and having them bid and awarded separately nearer the time of required delivery or installation. Some examples of these conditions are contained in Chapter 12.

6. The logistics of some multielement projects. Examples: In a new airport terminal complex, a new parking deck, road system, and outside electrical distribution system had to be completed before certain site areas were available so that the balance of the project could commence, and separate interior contracts were necessary for each of the airlines' interior and aircraft passenger loading units. Another example: In a municipal complex in a redevelopment area, demolition, parking decks, and street work in the project area had to be completed first. Then one building in the five-building complex had to be completed to demolish its predecessor that occupied part of the site. Others were started simultaneously with later contracts for graphics and landscaping.

It is usually in the owner's best interest to subsequently transfer an early or separately awarded contract into a general contract except in cases similar to 6 above. Even then, it is sometimes workable and therefore advisable. One reason for this is that companies usually found in the role of subcontractor will generally bid somewhat higher when there are doubts about coordination, access to work areas, protection of completed work, and intertrade scheduling. This can more than offset savings (if any) in general contractor markup in most situations. Further, there is much greater exposure for the owner to unjustified cost increases and contract time extensions when there is other than the simplest contract format.

It is therefore desirable to transfer early or other separately awarded contracts when

1. Work under the early or separately awarded contract would not be concluded prior to the commencement of the work under the general or other construction contracts and would have one or more interface activities requiring close coordination, and

2. The resulting general contract would not be so large as to limit the number of candidates for the general contract in the first place

Except in the most unusual circumstances, when both the above conditions exist, the construction manager or architect/engineer will want to plan on transferring the early or other separately awarded contracts into a general contract.

Planning for the transfer of the first contract to be issued for bidding or negotiation must begin prior to the completion of the contract documents.

Two fundamental approaches should be kept in mind in preparing documentation for transferable contracts. They are as follows:

1. Both the contractor who will become a subcontractor by the transfer and the contractor who will continue as a prime contractor, receiving the transferred contract into their contracts, must have had the maximum amount of information about the other contract, its status, and (if feasible) about the other contractor.

2. Transfer provisions in the special conditions should always be "transfer rights" of the owner, with no definite stipulation as to whether or not the contract will actually be transferred.

Figure 10-1 is a specimen special conditions section for a contract that may be transferred at or after the time of award into another contract, with the particular contractor in point then becoming a subcontractor to the other contractor.

Figure 10-2 is a specimen special condition provision for a general contract reserving rights for the owner to have several earlier awarded contracts transferred to the general contract, those contracts thereby becoming subcontracts.

The two specimen provisions illustrated in Figures 10-1 and 10-2 should be considered as examples of only two of a large number of provisions that might be utilized under similar conditions. It is very important to analyze each situation carefully and to write clearly in the special conditions of the contract documents to cover all aspects of the transfer, retaining all feasible options for the owner.

FIGURE 10-1

TRANSFERABLE CONTRACT

Division 1 / Section 1D

GENERAL REQUIREMENTS:

1. The Owner reserves the right to transfer this contract to the General Contractor who will perform the work under the base bid. Upon transfer, all of the Owner's rights and obligations under this contract shall be transferred and will flow directly to the General Contractor. After the transfer is accomplished, all payments to the Structural Steel Contractor, under his contract, shall become the obligation of the General Contractor. Likewise, all rights of the Owner under this contract, including the time of completion requirements, shall flow to the General Contractor.

SPECIFIC REQUIREMENTS:

2. As a condition to transfer of this contract, this Contractor must have provided a certificate of insurance for the specified amounts and the Owner must give this Structural Steel Contractor written notice five (5) days in advance of the effective date of transfer.
3. As a condition of transfer of this contract, any General Contractor to whom this contract is subsequently transferred shall have provided a 100 percent performance bond and labor and materials payment bond, AIA Document A311.
4. After transfer has been accomplished, neither the Owner nor this Structural Steel Contractor shall have any further direct obligation to one another, and this Contractor will have become a Subcontractor to the General Contractor and shall be known as the Structural Steel Subcontractor.
5. After work of this contract has been completed, the General Contractor, if the contract has been transferred, shall close out this contract and make all payments required by the contract but may withhold a maximum of 5 percent of the total Structural Steel Contract price until the work of this contract is accepted by the Owner.
6. Should the Owner elect to not transfer this contract, the Owner will close out this contract and make all payments required by this contract but may withhold a maximum of 10 percent of the total of this contract price until the work is accepted.

TRANSFERABLE CONTRACTS:

7. Separate contracts for segregated parts of the work which the Owner reserves the right to transfer are as follows:
8. Project Name: Medical Facility - Structural Steel
   Date of Documents: May 18, 1973
   Contractor: Steel Erectors, Inc.
   Address: 801 Industrial Way
   City: Norcross State: Georgia
   Amount of Contract: One hundred and fifty thousand Dollars ($150,000).

FIGURE 10-2

TRANSFERABLE CONTRACTS

Division 1 / Section 1E

GENERAL REQUIREMENTS:

1. The Owner has awarded separate contracts for segregated parts of the work as specified. The Owner reserves the right to transfer these contracts to the General Contractor, who will perform the work under the base bid. Upon transfer, all of the Owner's rights and obligations under transferable contracts shall be transferred to and will flow directly to the General Contractor. After the transfer is accomplished, all payments to the Contractor under the transferable contracts shall become the obligation of the General Contractor. Likewise, all rights of the Owner under transferable contracts, including Liquidated Damage provisions and the Time for Completion requirements, shall flow to the General Contractor.

SPECIFIC REQUIREMENTS:

2. As a condition to the transfer of contracts, the General Contractor must have provided a 100 percent performance bond and labor and material payment bond, AIA Document A311, and the Owner must give the Contractor of the transferable contract written notice five (5) days in advance of the effective date of transfer.
3. The General Contractor shall receive no compensation in addition to his contract price under the base bid by reason of receiving the transferred contracts.
4. After transfer has been accomplished, neither the Owner nor the Contractor of the transferable contract shall have any further direct obligation to one another, and the Contractor of the transferable contract will have become a subcontractor to the General Contractor.
5. After work transferred has been completed, the General Contractor, if the contract has been transferred, shall close out the transferred contract and make all payments required by the transferred contract, but may withhold a maximum of 5 percent of the total transferred contract price until the work transferred is accepted by the Owner.
6. Should the Owner elect to not transfer the contract, the Owner will close out the transferable contract and make all payments required by the contract, but he may withhold a maximum of 10 percent of the total of the contract price until the work is accepted.
7. All correspondence, change orders, shop drawings, and other data received by the Owner or Architect prior to the date of transfer shall be transmitted to the General Contractor after the date of actual transfer of contract.

Note that in Figure 10-2, the general contractor has been instructed to include in his base bid proposal sufficient sums to allow him to accept the transfer of the several "subcontracts" without any additional markup. In this case, the assumption of the construction manager or architect/engineer would have been that either transfer is a virtual certainty or the size of the contracts to be transferred and the cost caused the general contractor over and above the cost he would already have had in coordination responsibility was negligible.

A variation on the specimen provision illustrated in Figure 10-2 would be to instruct the general contractor, both in the special conditions provision and in a place in the proposal form, to quote on a percentage of the transferred contracts' prices for accepting them as subcontracts. In any case, however, the amount that most general contractors will usually add for accepting transferred contracts is relatively small. One reason for eliminating the extra charge provision in the proposal form is that this can be a complicating factor in determining the low bidder in the general contract bids in the first place. A further alternative is for there to be a specific stipulation in the transfer provisions of the general contract stating an amount (usually in a percentage) that will be added to the contract price of the other contracts that might be transferred into the general contract as a subcontract.

With the exception of the cost of the performance and payment bond, a reasonable allowance for general contractor's markup for transferred contracts, covering work that was already his responsibility to coordinate, would range from 1½ to 5 percent, depending on the size of the contract, the degree of interface, and the speed with which the owner would normally make progress payments. (The 5 percent upper limit above assumes that the owner would, in any case, be making progress payments within thirty days of individual monthly request certification by the construction manager or architect/engineer.)

With regard to the performance and payment bond in transferred contracts, there are several different approaches that may be taken. In most general contracting, bonded subcontractors are not nearly as prevalent as are bonds on contracts that flow directly to the owner. As a matter of fact, many small but reputable subcontractors will have some difficulty in obtaining a bond, particularly more specialized subcontractors that are not as frequently found with separate contracts as is an owner. These might include precast concrete subcontractors, ceiling systems subcontractors, smaller specialty structural frame contractors, etc. Therefore, when separate early contracts must be awarded to these types of contractors, it will be important for the construction manager or architect/engineer to be aware of these circumstances and to fully inform the owner. Second, he will need to determine some other method, formal or informal, of seeing to it that the bidding specialty contractors are in reasonable financial condition, able to do the work, and have a good reputation. In turn, when these contracts are to be subsequently transferred to a general contractor, it would probably be desirable to have the general contractor provide the owner with a 100 percent performance and payment bond covering the full amount of the general contract price posttransfer. This requirement will add about ½ to 1 percent to the amount of the markup the general contractor would be entitled to in the event of transfer. This situation should not present general contractors with any particular problem as long as a reasonable markup addition has been allowed, if they were in full possession of all information about the earlier or separately awarded contracts and of the contractors for that work at the time they placed their general contract proposal.

Another question related to bonds that must be dealt with is that of the bondability of specialty contractors. This problem must be considered in the case of major mechanical and electrical specialty contractors as well as larger contractors in the fields of grading, paving, foundations, and elevators. Unquestionably, this is a "cleaner" transfer in the eyes of the general contractor. However, if the 100 percent performance and payment bond is required of the general contractor, the fact that the subcontractors are fully bonded will not necessarily mean a reduction in the general contractor's bond premium cost, though it may reduce it somewhat in the case of very large contracts with general contractors of exceptionally good financial condition. However, at least ½ to ¾ of a percent markup should still probably be allowed, in most cases, for the general contractor when separately awarded bonded contracts are to be transferred into the general contract.

Unless the owner, such as a governmental subdivision owner, has the requirement for 100 percent bonding of the general contractor, there is probably no serious risk to the owner in having the general contractor bond be for an amount less than 100 percent of his posttransfer contract price as long as all contracts that are transferred into his contract are fully bonded. This would reduce the cost to the owner by a small amount and should not present him with any less surety as long as the requirements of the bond on the transferred "subcontractor" in the first place have required that the owner's rights under the bond remain the rights of the owner after any transfer of the contract into another contract.

In the case of early awarded contracts for constraint activity work or industrialized building systems work, about the only assurance that can be provided about the general contractor in the specialty contractor's transfer provision, as shown in Figure 10-1, is that the general contractor be fully bondable. This is certainly the case for governmental owners, since bidders cannot be limited in most instances. And while private owners might be able to select a list of general contract bidders, it would be unwise to make that commitment to the specialty contractors for early awarded contracts.

However, as far as the general contractor is concerned, for early awarded constraint activity contracts or industrialized building systems contracts, all information about the awarded contract and its contractor can and should be made available to the general contract bidders before they place their proposals. Each general contract bidder should have copies of the full agreements between the owner and the early awarded contractors, including the price, agreement, drawings and specifications, and all change orders to date. In the case of such early awarded work as grading, foundation, etc., it can and should be a requirement for the bidding general contractors to visually inspect the progress of the work as of a certain date and for all information to be made available to the bidders about the work that has been done.

Where layout is critical, as in the case of a foundation contract that is transferred later into a general contract, a good procedure is for the construction manager or architect/engineer to appoint an independent civil engineer surveyor (a properly licensed and certified surveyor) to do the layout work for the foundation contractor. In addition, the same surveyor would be identified in the subsequently awarded general contract as a professional subcontractor to the general contractor, and the general contractor would have to accept him with the placement of the bid. The general contractor in this and similar other instances, of course, would always have the right of access to the project and an opportunity to separately check the work. However, from a practical standpoint, a general contractor will generally employ such a surveyor to do layout work of this nature for coordination between his own subcontractors. Therefore, assuming that the construction manager or architect/engineer has exerted due care in selecting the surveyor, the general contractor is in precisely the same position, for all practical purposes, that he would be in had the foundation work been in his contract at the outset. The owner, in turn, would have accepted none of the risks of layout coordination problems in the transfer of the two contracts.

For separate contracts that are to be bid at or near the same time as the general contract, there are several important procedures to follow. This situation can come about for two reasons. One would be statutory conditions such as those that exist in states where mechanical and electrical work must be bid and awarded separately but may be subsequently transferred into a general contract. Another instance would be a desire on the part of the owner, construction manager, or architect/engineer to identify the major subcontractors, identify the exact price of their work, and be sure that these "subcontracts" are not "shopped" by the general contractor. (A general contractor "shops" when he continues to exert competitive forces and to obtain further prices on subcontracts after award of the general contract. This is not really an altogether serious concern of the construction manager or architect/engineer unless the general contractor in question has a reputation for ending up with poor quality subcontractors in this way or unless there is some very special requirement or qualification for the respective subcontractor. If necessary, the major subcontractors can be identified by having them named in a general contractor's bid.) However, it is difficult—if it needs to be known for accounting, professional fee calculation, or other purposes—to be absolutely certain of the exact price of subcontract by any other means other than a separate bid.

When either of the two foregoing conditions exists and separate bids for specialty work are to be obtained at the same time as the general contract bids, the following procedures are recommended:

The construction manager or architect/engineer should announce in the instructions to general contract bidders that no further bidding documents will be issued after a stipulated number of days preceding the date of receipt of bids. (Seven days might be reasonable in most instances.) This procedure has been found to be legal in most public projects, and of course there are no restrictions on this type of requirement in private projects. The day following the cutoff of issuance of documents, a list of all general contract bidders who have obtained bidding documents is issued to all specialty contract bidders, but without any claim as to whether or not these are all the general contract bidders there are or that all these will bid.

Meanwhile, the date for receipt of all separate specialty contracts, such as mechanical and electrical work, will have been set for twenty-four hours preceding the time for the receipt of the general contract bids.

All general contract bidders will be invited to attend the separate specialty contract bid opening. In the special conditions of the general contract documents, it will have been stated that the owner will have the right to award the contract to any specialty contract bidder and that it may be awarded for the amount of base bid or any amount more or less than the base bid up to a stipulated percentage and within a stipulated number of days thereafter. The reason for this is that the owner should reserve the option to rebid or negotiate in the event of bids exceeding estimates. In most projects, a 20 percent variation would be reasonable, and unless the separately bid specialty contract constituted a constraint activity, thirty to forty-five days would be reasonable for a time of rebid or negotiation.

By following the above procedure in the case of the two different types of contracts—i.e., the general contract and the separate specialty contracts —each bidder has a maximum amount of information available about the other contractors and prices and cannot later claim that a given general or subcontractor was unacceptable to him, and the owner has retained all reasonable options.

Of course, whenever a transfer serves no real useful purpose to the owner, it should certainly be avoided because of the complications that are inherent in such transfers. For example, if constraint work awarded early consists of access street and offsite utility work and it is feasible for this work to be completed prior to the bidding for the general contract, no purpose would be served by transferring this into the general contract, even if final payment has not been made to early awarded contract or even if the work were running a small degree behind schedule. In this instance, of course, one requirement on the general contractor should be that he visit the work and that he be required to "accept the work" by placing his bid for the general contract.

CHAPTER 11

Bid and Negotiations Management

The basic method of cost control in construction, and in particular the major method of cost savings, is the design process itself.

In order to plot the course of projected costs during the design process, the continuous and careful detailed estimating that has been discussed earlier (in the chapter on cost control) is employed, which in turn is based on utilizing market price information.

In turn, it is through competitive bidding or direct negotiation for construction contracts that the fair market prices must be obtained in order to attain the predictable results of the design and estimating efforts. Many carefully designed buildings that have been carefully estimated during the design process and for which complete and well-prepared contract documents have been issued have ended up with poor prices due to a lack of management of and planning for the bidding and negotiation phase.

The management of the bidding and negotiation phase must commence long before completion of the construction contract documents. The first order of business is, of course, to determine the posture of the owner for purchasing construction. Is the owner of group I or group II? The characteristics of owners relative to the responsiveness of the construction industry and the subsequent categorizing into groups I and II were discussed earlier in Chapter 3. As was discussed in that chapter, the construction manager or architect/engineer will generally want to obtain competitive bids for group I owners and give consideration to direct negotiation for group II owners. One additional point that might be mentioned here, regarding responsiveness of the construction industry, is that it is possible for a sizable architect/engineer firm, as well as a construction management firm, to have a sufficient volume of business itself to begin to develop group II owner characteristics for its clients, whether or not its clients themselves would ordinarily be considered members of the group II category. However, this conclusion should be arrived at cautiously and can sometimes easily be overestimated.

Since the management of bidding is substantially different from the management of negotiations, this chapter will be divided into two major subdivisions, the first dealing with the management of bidding and the second dealing with the management of direct negotiation.

MANAGEMENT OF BIDDING

The bywords of bidding in the construction industry are "competition" and clarity so that the competition will subsequently provide the right price, one that will hold up well once the contract has been executed.
present with competitive pressures at work that will obtain fair market prices. Secondly, he must be certain that this competition takes place with optimum

clarity so that the competition will subsequently provide the right price, one that will hold up well once the contract has been executed.

At least several weeks prior to the date of issue of the contract documents, the construction manager or architect/engineer should carefully identify prospective bidders. This should be the case whether the project is a public project or private project. In the case of the public project, in most political subdivisions the governmental owner cannot limit (except by contractor prequalification in some instances) the bidders on a project. Private work, of course, is different in that bidders can be selected, which is a very good procedure to follow.

In either case, however, it is important to identify sufficient competitors who are "right" for the job. The term "right" refers to a contractor who is of an appropriate size, has an appropriate proximity to the project, has experience in the building type, and is looking for new contracts. Other characteristics would be that of having a good reputation in the industry, both with previous owners and suppliers/subcontractors, and being known to consistently bid competitively for contracts.

Upon the identification of a prospective bidder, a good procedure is to actually meet with the prospective bidder and discuss the project. As long as there is to be normal competition at work, there would seem to be no reason not to let a prospective bidder know the amount of the owner's budget or estimate for a project. Before issuing the bidding documents themselves, it is a good procedure to invite prospective bidders to the architect/engineer's office to review renderings, models, preliminary drawings, working drawings in process, etc.

After a discussion with each prospective bidder of the schedule and other salient points about the project, it should be determined that the bidder will be available for the scheduled dates of document issue, prebid conference, and receipt of bids. It will be important, during this time, for the construction manager or architect/engineer to evaluate for himself the degree of interest, and be convinced that the prospective bidder is seriously interested, has scheduled the required time in his estimating department, and has scheduled the dates of issue, prebid conference, and bid.

This same process should be followed for all prospective bidders and should be concluded shortly before the scheduled issue date, rather than too far in advance, so that the respective prospective bidder situation will be less likely to change.

Five or six serious, "right," and committed prospective bidders would generally be considered sufficient for most projects. Except under unusual circumstances, having more than six or seven bidders is not usually in the best interests of the owner. The reason for this is that when there is an excessive number of bidders, each bidder, in evaluating his own situation, must consider that he has a relatively small chance for having the project awarded to him. Therefore he is less likely to put in the extensive effort that is necessary to bid a job competitively and to give it the careful scheduling and attention that it needs. It would not be unusual to find a nine- or ten-contractor bidding list dwindle to only three or four bidders by the date of receipt of bids, while a five- or six-contractor list may hold without any shrinkage during the bid period.

After the prospective bidders have been well identified, the next step will be the issue of the bidding documents. It is important to follow the issuance of these documents carefully to be sure that each of the prospective bidders has, in fact, requested the documents and to see, in turn, that the documents are properly issued and received by the prospective bidder. This kind of quick follow-up is very helpful in making an early determination that the prospective competition that was earlier evaluated is really there and to avoid any confusion on the part of bidders relative to the issue of documents.

During this same period, legal advertisements will be placed, if they have not been placed prior to this time, (using legal journals in the case of public work). In the cases of both private and public projects, advance notice should be given to the various construction reporting services, and in the case of very large projects, to construction periodicals such as the *Engineering News-Record*.

Also at the time of document issue, the construction manager or architect/engineer will want to see that there are one or more sets of documents in all plan rooms that are likely to be used by general contractor, subcontractors, and suppliers.

In the document issue and deposit procedure of the construction manager or architect/engineer, consideration should be given to some reasonable means of making sufficient bidding documents available to suppliers, manufacturers, and subcontractors. However, the practice of issuing partial documents for any purpose whatsoever should be avoided.

In the preparation of the legal advertisement in the case of public projects, in the preparation of informal notices in the case of private projects, and in the preparation of the instructions to bidders that are bound into the specifications book, the time and place of the prebid conference should be set down. See Figure 11-1 for a specimen legal advertisement and Figure 11-2 for a specimen instructions to bidders, the latter as it would appear as part of the contract specifications.

FIGURE 11-1

INVITATION TO BID

Division A / Section A1

Sealed proposals are invited for the construction of a Medical Facility Building to be erected on the property of the Owner located at Atlanta Medical College in the city of Atlanta, state of Georgia, and as further described by documents prepared by Heery & Heery, Architects and Engineers, of 880 West Peachtree Street, N.W., Atlanta, Georgia 30309.

A Prebid conference is scheduled for 2:00 P.M., time legally prevailing, November 8, 1973, at the office of the Architect. All prospective bidders are required to attend this conference and all prospective subcontractors and materials vendors are invited to attend.

Proposals will be received at Room 200, Bleckley Hall, Atlanta Medical College, until 2:00 P.M., time legally prevailing, on the 27th day of November, 1973, at which time and place all proposals received will be opened and read aloud. Proposals received after time is called will be returned.

Bidding documents may be examined in the office of the Architects and Engineers, the F. W. Dodge Plan Rooms in Atlanta, Georgia; the Builders' Exchange Plan Rooms in Atlanta, Georgia; the Scan Plan Service in Atlanta, Georgia.

Base-bid Contractors may obtain copies of bidding documents from the office of the Architect upon deposit of $100.00 per set. Deposit on as many as two (2) sets of documents will be refunded upon submission of a bona fide bid and return of complete documents in good condition within twenty (20) days following the opening of bids. All other deposits will be refunded with deductions approximating cost of reproduction of documents upon return of same in good condition within thirty (30) days after date of opening of bids.

If awarded, contract shall be on a lump-sum basis. Proposal may not be withdrawn for a period of thirty-five (35) days after time has been called on the day of opening, without consent of the Owner. A bid bond in the amount of 5 percent of the base bid will be required.

The Owner reserves the right to reject any and all bids, to waive any formalities in bidding, and to award a contract for any part of the work, or the job as a whole.

OWNER: ____________________

ADDRESS: ____________________

CITY: ____________________ STATE: __________

FIGURE 11-2

INSTRUCTIONS TO BIDDERS

Division A / Section A2

GENERAL INFORMATION:

1. Refer to Invitation to Bid for information relating to time, date, and place for receipt of proposals and other pertinent bidding information.

2. Refer to Special Conditions for information relating to time for completion, liquidated damages, substitution of materials and other special conditions pertinent to work.

3. Refer to General Conditions and to Supplementary Conditions for information relating to payments, guaranty bonds, taxes, insurance, and other conditions pertinent to the work.

EXAMINATION OF BIDDING DOCUMENTS AND ADDENDA:

4. Each Bidder shall carefully examine bidding documents and all addenda and thoroughly familiarize himself with the detailed requirements prior to submitting a proposal. Should a bidder find discrepancies or ambiguities in or omission from bidding documents, or should he be in doubt as to their meaning, he shall at once and, in any event not later than seven (7) days prior to bid date, notify the Architect, who will send written addenda to all bidders. The Architect will not be responsible for any oral instructions. All addenda sent to bidders will become a part of contract documents. All inquiries shall be directed to Architect's office, Heery & Heery, 880 West Peachtree Street, N. W. Atlanta, Georgia 30309, Telephone 881-9880, Area Code 404. No allowance will be made after bids are received for oversight by Bidder.

5. Changes or corrections may be made in the bidding documents after they have been issued and before bids are received. In such case a written addendum describing the change or correction will be issued by the Architect to all bidders. Such addendum or addenda shall take precedence over the portion of the bidding documents concerned and shall be considered as part of the bidding documents. Except in unusual cases, an addendum will be issued to reach the Bidder at least three days prior to the time set for receiving bids. Bidder shall acknowledge receipt of all addenda on the proposal form in the space provided.

EXAMINATION OF CONDITIONS AFFECTING WORK:

6. Prior to submitting a proposal, each Bidder shall examine and thoroughly familiarize himself with all existing conditions including all applicable laws, codes, ordinances, rules and regulations that will affect his work. Bidders shall visit the site, examine the grounds and all existing buildings, utilities, and roads, and shall ascertain by any reasonable means all conditions that will in any manner affect work. Bidders shall ask the Architect for any additional information deemed necessary for them to be fully informed as to exactly what is to be expected prior to submitting a proposal. The drawings have been prepared on the basis of surveys and inspections of the site and represent an essentially accurate indication of the physical conditions at the site. This, however, shall not relieve the Bidder of the necessity for fully informing himself as to existing physical conditions.

(Continued)

PREPARATION OF PROPOSALS:

7. To be considered as eligible to submit a proposal, a Bidder must be legally licensed to operate under applicable laws of the state in which the work is to be executed. Envelopes containing proposals shall be opaque, and must be so presented that they may be easily identified as containing a proposal. Outside of envelope must show:

(a) Name and location of project as described in the bidding documents
(b) Name and address of Bidder
(c) Bidder's contractor's license number, if applicable
(d) Bidder's bidding license number, if applicable
(e) Identification of work for which proposal is submitted, i.e., general contract, etc.

8. Submit proposal on forms furnished by the Architect. Oral or telephonic proposals or modifications will not be considered. Telegraphic bids will not be considered, but modification by telegraph of bids already submitted will be considered if received prior to the time set for receiving proposals; telegraphic modifications shall not reveal the amount of the original or revised proposal. All blank spaces on the form must be filled in. Signature must be in longhand and be executed by a principal duly authorized to make contracts. Bidder's legal name must be fully stated. Completed form must be without interlineation, alteration, or erasure. No proposals will be considered after calling of time, regardless of how they are transmitted. Proposals shall not contain any added statement that will recapitulate, modify, or interpret the terms of the proposal.

9. Each proposal must be accompanied by a bid bond in the amount of five percent (5%) of the base bid, made payable unconditionally to the owner. This is required as evidence of good faith and as a guarantee that, if awarded the contract, Bidder will execute contract and furnish required bonds and evidence of insurance within ten (10) days after receipt of Notice of Acceptance. Proposal without bid bond will not be considered.

10. Bid bonds will be returned within 48 hours after Owner and successful Contractor have executed contract and executed performance bond has been approved by Owner. If no award has been made within thirty-five (35) days after opening of proposals, upon demand of a Bidder at any time thereafter, bid bonds or checks will be returned provided that he has not been notified of acceptance of his proposal.

11. Proposal may not be withdrawn for a period of thirty-five (35) days after time has been called on the date of opening except by mutual consent of owner and Bidder and except that proposals may be withdrawn on written or telegraphic request received from Bidders prior to time fixed for receiving proposals. Negligence on the part of Bidders in preparing proposals confers no right for the withdrawal of proposals after opening.

ACCEPTANCE OR REJECTION OF PROPOSALS:

12. The Owner reserves right to reject any and all proposals when such rejection is in the interest of Owner; to reject proposal of a Bidder who has previously failed to perform properly or complete on time contracts of a similar nature; and to reject proposal of a Bidder who is not, in opinion of Architect or Owner, in a position to perform the contract. Owner also reserves right to waive any informalities and

technicalities in bidding. Owner may also accept or reject any of the alternates set forth. Contract will be awarded (unless all bids are rejected), under normal circumstances, to lowest responsible Bidder. Owner reserves right, however, to award contract to serve his best interest, and therefore may select a Bidder other than the lowest.

BIDDER'S QUALIFICATIONS:

13. Before a proposal is considered for award, the Bidder may be requested by the Architect to submit a statement of facts in detail as to his previous experience in performing similar or comparable work, his business and technical organization, his financial resources and the plant to be used in performing contemplated work

DRAWINGS AND SPECIFICATIONS:

14. All documents furnished to any person, under any condition, remain property of the Architect and shall immediately be returned upon request and, in any case, not later than twenty (20) days after receipt of proposals. All checks for documents shall be made payable to the Architect.

15. Bidding documents will be placed on file and may be examined during normal office hours at:

(a) Architect's Office in Atlanta, Georgia
(b) The F. W. Dodge Plan Rooms in Atlanta, Georgia
(c) The Builders' Exchange Offices in Atlanta, Georgia
(d) The Scan Plan Service in Atlanta, Georgia

16. Base bid contractors may obtain copies of bidding documents from the office of the Architect upon deposit of $100.00 per set. Deposit on as many as two (2) sets of documents will be refunded upon submission of a bona fide bid and return of complete documents in good condition within twenty (20) days following the opening of bids.

17. Additional copies of bidding documents may be obtained by base bid contractors upon payment of $100.00 per set. These documents shall remain property of the Architect, and if returned in good condition, complete, within twenty (20) days after bids are received, the payment less cost of reproduction will be refunded.

18. Sub-bidders and vendors may obtain copies of bidding documents from the office of the Architect upon deposit of $100.00 per set. Bidding documents remain property of the Architect, and upon return of complete documents in good condition, the payment less cost of reproduction will be refunded.

SUBSTITUTIONS:

19. The attention of Bidders and all other parties is called to the conditions set forth in the Special Conditions section of these specifications.

SCHEDULES AND REPORTS:

20. The attention of Bidders and all other parties is called to the conditions of the Schedules and Reports section of these specifications.

PREBID CONFERENCE

The prebid conference has several purposes and is a procedure that should be employed in all contracts, whether they are general or part of a series of multiple contracts. The purposes of the prebid conference are as follows:

1. To make a midway evaluation of competitive interests
2. To brief prospective bidders and other interested parties, such as subcontractors and suppliers, on the project schedule, usually in the form of a critical path network
3. To brief prospective bidders on CPM fundamentals in general and to make appointments for each bidder with the CPM consultant for private consultation and provisional preliminary network modification prior to the date of bid receipt
4. To bring out questions regarding the contract documents early in the bid phase so as to avoid extensive last-minute addenda

Most contractors of any size who participate in the competitive bidding market are consistently bidding projects. In turn, their situation frequently changes in any given period of four to six weeks. The prebid conference, if properly employed, will tell the construction manager or architect/engineer a good bit about continuing interest in the project on the part of each prospective bidder. The date for the conference should usually be set about halfway through the bidding period. Typically, this will be two to three weeks, and in some cases longer, before the date of receipt of bids. In the case of a serious shrinkage in bidders, this may even allow time to replace one or two lost bidders, or—if there is not time to replace the bidder in the remaining time—to reset the scheduled date for receipt of bids to allow adequate time to attract new bidders. In no case should the construction manager or architect/engineer allow bidding to go forward to the date of receipt of bids knowing that there are less than three serious, active, and "right" bidders for the project.

Prebid conference in session.

As was said earlier, "clarity" is one of the two major bywords of bid management. Many times good bidding is undermined at the last minute by extensive and complex addenda or by the presence of too many unanswered questions in the contract documents, so that only hours before the bids are to go into the owner the contractor is put into the position of having to unrealistically increase his price simply to protect himself. This is not only very detrimental to the owner but is most disheartening to a bidding contractor who has put in a great deal of time, effort, and expense only to reach the point of bid placement and find that he is in an unclear position and cannot price sharply.

(Of course, the major aspect of clarity is the completeness, accuracy, and form of the contract drawings and specifications themselves. This major activity of the architect/engineer which has gone on before the actual time of bidding is crucial, to say the least, in receiving competitive bids or in direct negotiating for that matter. These documents must be exactly what their name indicates, construction "contract documents" that are binding on the parties to the agreement.)

The instructions to bidders should require the attendance at the prebid conference of all bidders. In addition, subcontractors, material suppliers, and other interested potential participants in the construction should all be invited to attend. This should be done through the medium of the instructions to bidders as well as in the reporting services and plan rooms.

The prebid conference should be conducted by the project manager of the construction management firm or the project manager of the architect/engineer if the architect/engineer is providing construction management services. In any case, however, key architects and engineers from the architect/engineer organizations as well as the CPM consultant should all be in attendance.

Minutes should be kept of the meeting, but only for the construction manager's or architect/engineer's internal purposes. They should not be forwarded to bidders. Any actual changes or clarifications in the contract documents should be in the form of an addendum issued after the prebid conference. This addendum should be forwarded promptly after the prebid conference.

With regard to CPM scheduling and the CPM consultant's activities in the prebid conference, see Chapter 9.

It is important to maintain contact with all prospective bidders on a regular basis throughout the bid period. One sign to watch for is a lack of questions from any one bidder. A bidder who asks no questions is probably not figuring the job. It is also important to keep tabs on major subcontract bidders for such portions of the work, in the case of a general contract, as the mechanical work, electrical work, structural framing, foundations, etc. Care must be taken to generate interest from major subcontractors in much the same way that competitive interest is generated prior to the issuance of the documents for general contractors.

Construction managers and architect/engineers should also be aware that certain trade associations tend to stifle competition. It is important to know in what areas of the country such associations might attempt to do so. When that situation occurs, it is almost always possible to attract the interest of bidders who are not members of that association. This will be crucial to obtaining fair market prices, and at least a couple of other bidders outside such an association should be brought into the bidding picture. Again, a subcontractor who is not asking questions is probably not figuring the job.

Long before the contract documents were completed and issued for bidding, the proposal form itself should have been worked out—in conjunction with the construction management plan—and approved by the owner. Actually, it is a good idea for the architect/engineer or construction manager to prepare this form as part of the documentation of the design development phase of document preparation. It should again be reviewed, in light of final quantity survey estimates, just before the contract documents are issued for bidding. Subsequently, it should be the subject of a detailed conference with the owner prior to the receipt of bids.

It is important for the construction manager or the architect/engineer to prepare the owner for receipt of bids. For typical projects, a good architect/engineer with an integrated design and estimating operation, exerting due care, should be able to give most owners (with the exception of some governmental subdivisions) a good assurance that contracts can be obtained for the owner within the architect/engineer's estimates. However, one can never be sure that a given set of bids will be within a given estimate. Consequently, the owner must understand this and at the same time know what procedures the construction manager or architect/engineer intends to follow in the case of bid overage.

A competent construction manager or architect/engineer must not let an owner, through misunderstanding, take erratic actions upon the opening of bids that are over the estimate. The best way to avoid this, other than having a good estimate and a carefully managed bidding, is to have the owner well prepared for that possibility, making sure that he understands the various procedures that the construction manager or architect/engineer can and will follow at that time.

One corrective measure that can be taken is the utilization of additive or deductive alternates in the proposals. Alternates should not be used indiscriminately, as an excessive number of alternates can cause sufficient confusion to actually adversely affect the base bid price itself, not to mention the prices of the alternates. See Figure 11-3 for a specimen proposal form including several alternates and unit prices.

FIGURE 11-3

FORM OF PROPOSAL

Division A / Section A3

TIME ____________ DATE ________, 19 ___

REGISTERED MAIL

TO: ______________________________ OWNER

______________________________ ADRESS

____

______________________________ CITY/STATE

FROM: ______________________________ BIDDER

______________________________ ADDRESS

______________________________ CITY/STATE

Operating as (strike out conditions that do not apply) an individual,a Company, a corporation, organized and existing under the law of the State of ____________, or a Proprietorship, a Partnership, or Joint Venture consisting of ______________

____ .

BASE BID PROPOSAL:

1. Having become completely familiar with the local conditions affecting the cost of work at the place where work is to be executed, and having carefully examined the site conditions as they currently exist, and having carefully examined Bidding Documents prepared by Heery & Heery, Architects and Engineers, and titled

____

Architects' Commission No. ______, dated ___________ together with any addenda to such Bidding Documents as listed hereinafter, the undersigned hereby proposes and agrees to provide all labor, materials, plant, equipment, transportation and other facilities as necessary and/or required to execute all of the work described by the aforesaid Bidding Documents for the lump-sum consideration of:

______________________________ Dollars ($ _________),
said amount being hereinafter referred to as the Base Bid or Base Bid Proposal.

(Continued)

ALTERNATES:

2. When alternate proposals are required by Bidding Documents, the undesigned proposes to perform alternates for stated resulting additions to or deductions from the Base Bid. Additions and deductions shall include any modifications or work or additional work that undersigned may be required to perform by reason of the acceptance of any alternate. (Note: Include all alternates as required by Bidding Documents).

ALTERNATE NO. 1:

Contemplates eliminating all automatic door operator equipment. Adjust Base Bid by ADDING or DEDUCTING (strike one)

______________________________ Dollars ($ __________) .

ALTERNATE NO. 2:

Contemplates eliminating all carpet and substituting vinyl asbestos tile in lieu therefor. Adjust Base Bid by ADDING or DEDUCTING (strike one)

______________________________ Dollars ($ __________) .

ALTERNATE NO. 3:

Contemplates deleting all pressure grouting and related foundation work and substituting precast concrete piles and their related foundation work. Adjust base bid by ADDING or DEDUCTING (strike one)

______________________________ Dollars ($ __________) .

ADDENDA ACKNOWLEDGEMENT:

3. The undersigned acknowledges receipt of the following addenda: (List by number and date appearing on addenda.)

Addendum No.	*Date*	*Addendum No.*	*Date*
________	________	________	________
________	________	________	________
________	________	________	________

TIME OF COMPLETION:

4. The undersigned agrees to complete the work within the number of calendar days as stipulated in the Special Conditions section of these specifications and to meet the specific dates set forth therein.

CHANGES IN WORK:

5. The undersigned agrees that when changes in work are ordered which involve extra cost over and above contract sum, and when such work, due to an emergency, is ordered to proceed on basis of cost-plus-fee, such fee shall be as required by the General Conditions and Supplementary General Conditions.

UNIT PRICES:

6. Unit prices are complete for labor, equipment, and material; overhead and profit for additions will be based on stated percentages.

Description		
Precast concrete piles	$ ______	per lineal foot
Rock removal	$ ______	per cubic foot

BID SECURITY:

7. Bid security in the amount of 5 percent of the Base Bid is attached, without endorsement, in the sum of

______________________________ Dollars ($__________)

which is to become the property of the Owner in the event the Contract and Performance and Payment Bonds are not executed within the time set forth as liquidated damages for the delay and additional work caused the Owner.

8. The undersigned agrees that upon receipt of the Notice of Acceptance of his bid, he will execute the formal Contract and will deliver a Surety Bond for the faithful performance of this Contract and such other bonds and insurance as may be required by the specifications.

9. The undersigned further agrees to execute the formal Contract within ten (10) days from date of notification of the acceptance of this Proposal, and, in case the undersigned fails or neglects to appear within the specified time, to execute the contract of which this Proposal, the Bidding Documents, and the Contract Documents are a part, the undersigned will be considered as having abandoned the Contract, and the Bidder's Bond accompanying this Proposal will be forfeited to the Owner by reason of such failure on the part of the undersigned.

10. The undersigned further agrees that the bid security may be retained by the Owner and that said proposal guaranty shall remain with the owner until the Contract has been signed and Performance Bond has been made and delivered to the Owner.

GENERAL STATEMENT:

11. The undersigned has checked all of the above figures and understands that Owner will not be responsible for any errors or omissions on part undersigned in preparing this Proposal.

12. In submitting this Proposal, it is understood that the right is reserved by Owner to reject any or all bids and waive all informalities in connection therewith. It is agreed that this Proposal may not be withdrawn for a period of thirty-five (35) days from time of opening.

(Continued)

13. The undersigned declares that the person or persons signing this Proposal is/are fully authorized to sign on behalf of the firm listed and to fully bind the firm listed to all the Proposal's conditions and provisions thereof.

14. It is agreed that no person or persons or company other than the firm listed below or as otherwise indicated has any interest whatsoever in this Proposal or the contract that may be entered into as a result of this Proposal and that in all respects the proposal is legal and firm, submitted in good faith without collusion or fraud.

15. It is agreed that the undersigned has complied or will comply with all requirements of local, state, and national laws, and that no legal requirement has been or will be violated in making or accepting this Proposal, in awarding the contract to him, and/or in the prosecution of the work required.

16. This Proposal has been prepared using sub-bids received from the firms listed below:

Classification of Work	Name of Sub-bidder	License Number
Electrical	______________	______________
Plumbing	______________	______________
Heating, Ventilating and Air Conditioning	______________	______________

Respectfully submitted, this

____ day of ______________, 19 ____

(Firm Name) ______________________

(Address) ______________________

____

(Signature) ______________________ L.S.

(Name Typed) ______________________

(Title) ______________________

(SEAL IF BIDDER IS A CORPORATION)

17. The following are names, titles, and addresses of the Proprietor, all Partners, or three Corporate Officers:

(LIST)

18. The following bank reference is given:

Name of Bank: ______________________

Address: ______________________

____

Officer of Bank: ______________________

ENCLOSURES:

1. Bid Bond
2. Statement of Bidder's Qualifications

COMM. NO.

STATEMENT OF BIDDERS QUALIFICATIONS
(Enclosure with Proposal)

To accompany bids submitted for construction of a ______________

______________ building at ______________

Name of bidder ______________________

Business address ______________________

When organized ______________________

Where incorporated ______________________

How many years have you been engaged in the contracting business under the present firm name? ______________

Financial statement ______________________

Credit available for this contract $ ______________

Contracts now in hand, gross amount $ ______________

Plan of organization ______________________

____

Have you ever refused to sign a contract at your original bid? ______________

(Continued)

Have you ever been declared in default on a contract? ________________

Remarks: ____

____

(The above statements must be subscribed and sworn to before a Notary Public).

Date ________________

Firm Name ________________

By ________________

Title ________________

Notary Public ____

Unit prices have been a traditional part of construction bidding but can be quite treacherous for an owner. The most common unit prices have been those for the removal of rock, additional excavation, or additional concrete in footings. All three of these should be avoided and are unnecessary in most proposals. The owner is almost always better off to have this work done on a force account basis (accounting for labor forces and materials after the work has been done by the contractor, in accordance with change order pricing procedures set out in the special conditions), because very little attention can be or is given the amount of such unit prices in the award of the contract itself. They are almost invariably unreasonable and the owner has his hands tied. See the discussion on change order pricing in Chapter 7.

In obtaining unit prices in the proposal for such things as pile lengths and caisson lengths, the following procedure should be followed: First, a very reasonable quantity of the piling lengths or caisson lengths, as the case may be, should be included in the base bid. It is important to have the included quantities set as accurately as possible so that it would be almost impossible for a bidder to discern whether or not there is likely to be an increasing or decreasing adjustment in piling or caisson length. In turn, unit prices should be obtained that are applicable as both additive and deductive adjustments (with a maximum deductive adjustment, such as ±30 percent, of the quantity involved) as a requirement in the proposal form. This way, the owner will receive the fair price and the contractor will not be penalized when these adjustments take place.

In other cases, unit prices can actually be stipulated in the proposal form.

This is a fairly unusual procedure, but it is perfectly legitimate in the proper circumstances. For example, in an industrial project where there is the likelihood of additional railroad lead track, with this being a fairly complicated force account pricing procedure, the construction manager or architect/engineer can usually learn quite accurately what is a fair price, on a lineal foot basis, for additional track to be laid. In turn, this price can be stipulated as a unit price in the proposal form, with a reasonable maximum being set.

Some construction contracts. primarily grading and paving, are almost altogether based on unit prices. However, the owner's safeguard in these instances is that the total base bid, as in the case of the British quantity survey method of "tenders," is a summation of the unit prices multiplied by the quantities that are the basis of the proposal. Unit prices are also discussed in Chapter 7.

The place of receipt and opening of bids should generally be set at a point most convenient to the largest number of bidding contractors so that they may use their normal office and work staff as close to bid placement time as possible.

At the opening, the construction manager or architect/engineer, with the permission of the owner, would do well to waive certain formalities, as outlined below.

He should announce that unless there is an objection from one of the bidders, all bids will be opened and reviewed for completeness. If it is noted that there has been a minor omission—for example, if a contractor still has the bid bond in his pocket by mistake, has failed to fill in a blank in the proposal form, or has failed to place a signature at some point on the documents—this type of inadvertent incompleteness will be allowed to be corrected, in the opening room, in the presence of all in attendance, before any of the bids are read aloud. The bids should then be read aloud in the order in which they have been opened.

A public opening is advisable for both private and public works. If a private owner has some objections to a completely public opening, he should be asked to at least agree to an opening that has one representative of each bidder present, with the understanding that the opening will be limited to those present and that no information will be given out except through the owner. However, an owner who asks a construction manager or architect to open bids in private is causing a great loss of credibility between the construction industry and the respective construction manager or architect/engineer as well as the owner. Therefore, since this will hurt both this owner and the construction manager and architect/engineer in the future project pricing, it should be avoided if at all possible.

Unless there is an extreme shortage of time and it is readily apparent that there is an acceptable bid placed at the time of the opening, it should be standard procedure to announce before the opening that "This is an opening only and there will be no announcement as to apparent low bidder or award at this time." In those rare circumstances where there is to be an immediate award, even then it should be announced that there will be a break after the opening and that the award will take place in a private meeting with the successful bidder after the owner and the construction manager or architect/engineer have had a chance to make a final review of bids and consult in private.

As soon as an award has been made, all bidders should be so notified and plan deposits should be promptly returned, with appropriate adjustments, along with the receipt of documents from the unsuccessful bidders. A personal call or note of thanks to the unsuccessful bidders from the construction manager or architect/engineer is always in order.

Should the low bid exceed the owner's budget and should acceptable deductive alternates not provide the basis for an immediate award, there are several courses fairly normally followed. These are listed below in order of preference:

1. Direct negotiation with the low bidder.
2. Preparation of an addendum and new proposal form with immediate rebidding, sometimes taking place within a matter of ten days to three weeks, sometimes limited to the low three bidders in the case of private work.
3. Separation from the major part of the work of some complicating portion that may be discovered as being the cost overrun problem. For example, if it were found that the electrical system was substantially over the budget and that all other work was within the budget, and if it is feasible for the rest of the work to proceed for, say, thirty or forty days before the first electrical work must be started, an award could be made to the low general contract bidder by direct negotiation or by quick rebid of the project without the electrical work. In turn, the electrical work redesign could commence concurrently and be bid at a later time in such a way that it could either continue as a freestanding contract or be transferred into the general contract.
4. Redesign and reissue for bidding, handled either on an addendum basis or a complete reissue basis, if the cost of doing so is justified or necessary.

In the contract document preparation phase, during the preparation of the contract time provisions, it should be stipulated that the contract time will commence either upon the execution of the construction contract between the owner and contractor or upon the issuance of a notice to proceed, whichever comes first. In turn, once an acceptable proposal is in hand and the owner is prepared to award the contract, if time is of the essence and if there should be any procedural or legal delays in the execution of the agreement itself, the construction manager or architect/engineer should issue the notice to proceed on behalf of the owner, stipulating therein the effective date of the notice to proceed. This should only be done, however, with written confirmation to him from the owner. See Figure 11-4 for a specimen notice to proceed letter. Of course, it is always preferable to have the contract itself executed, and it would be best if the construction manager or architect/engineer could have the owner prepared to do so at the time of scheduled award.

FIGURE 11-4

SPECIMEN NOTICE TO PROCEED

Atlanta
January 6, 1973

Rizzo Construction Company
Post Office Box 1753
Atlanta, Georgia 30318

Attention: Mr. John P. Smith

Re: Manufacturing Facility for ABM Corporation

Gentlemen:

The Owner has authorized us to notify you that you are to proceed as of January 6,1973 with the work in accordance with drawings and specifications, the Owner's acceptance of proposal of $4,743,639.00, less agreed upon deductions (see "Exhibit A"), giving a final Contract Award price of $4,637,906.00.

The contract will be forwarded to you for your execution shortly and the specific dates of Section 1A will reflect the twelve-day delay of this notice to proceed.

Very truly yours,

HEERY & HEERY

James D. Atwood

JDA:mm

cc: Mr. I. C. Elroy
7299a

MANAGEMENT OF NEGOTIATION

As competitive bids are generally received for group I owners, direct negotiation should generally be given first consideration for group II owners. There can be some substantial advantages to the group II owner in direct negotiation when those conditions of responsiveness that characterize the group II owner are present. One misconception, however, regarding negotiation is that it saves time in the commencement of construction. This is frequently not the case. If the need exists to commence a constraint activity early, such as the award of the structural frame contract, this can be dealt with on either a competitive bid or negotiated basis in approximately the same time frame in accordance with the methods and procedures discussed in Chapter 10.

In the award of a single lump-sum contract on a negotiated basis or in the case of the award of the major general construction contract through direct negotiation, the contractor must go through very nearly the same procedures, requiring about the same and in some cases longer periods of time to confirm his final prices with subcontractors, suppliers, and others. This can only be done, precédent to executing a contract with the owner, after complete drawings and specifications are in hand. While it is true that an "upset price" might be obtained prior to this time, the real contract cannot really be in hand until the documentation is complete and prices have been confirmed.

Therefore it is necessary to allow approximately the same amount of time after completion of contract documents for final confirmation of price and

Teaching hospital, 265 beds, for Medical College of Georgia. (Heery & Heery, Architects & Engineers.)

execution of the contract in the case of a negotiated contract as is required for bidding and award in the competitive bid case. Further, there is always the possibility that the negotiations will not be successful, requiring then a further time period for negotiation with another contractor or for competitive bidding. Therefore the same contingency time that might be required in the case of rebid for group I owners really has to be provided for in scheduling negotiated contracts for group II owners.

The major advantages of utilizing a negotiated contract are as follows:

1. It makes possible the direct selection of a contractor who has a good reputation, is in good financial condition, is best suited to handle the project, and is known to be responsive to the owner.

2. A maximum or "upset price" and confirmed time schedule can be obtained from a contractor earlier in the design/construction sequence.

3. Advice on construction methods, sequencing, materials, and other cost-related matters can be obtained earlier in the design/construction sequence from the contractor.

In analyzing the pros and cons of going forward with a negotiated contract in a given project, the construction manager and architect/engineer should be careful not to overestimate the value of items 2 and 3 above. These are popular conceptions and do, in fact, often take place. However, they do not *always* take place, and the construction manager or architect/engineer can be greatly disappointed to find that the kind of information available from the contractor was either of no help or was readily available to him from other sources. He can also be greatly disappointed to find that once the contractor who has been selected for negotiation feels that he has the contract, schedules and cost goals along with general condition provisions become much less ambitious and much less favorable to the owner.

Therefore there are two very important procedures that should be followed from the beginning of the schematic design phase on through contract award in the case of projects for which negotiation is intended. First, the construction manager or architect/engineer should carry on precisely the same full estimating program in conjunction with the design effort, independent from cost estimating and information received from the selected contractor, just as he would do in a project that is to be competitively bid. The information received then from the internal estimator should be constantly compared with the cost data that is being given by the contractor. By the time the construction contract documents are complete, the internal estimator should have come up with his own, completely independent quantity survey estimate, which should then be used by the construction manager and architect/engineer with the owner in making a judgment as to the award of the negotiated contract.

Second, the general and special conditions of the proposed contract should generally follow the same format and include the same kinds of provisions for a negotiated contract that they would include for a bid contract. These general and special conditions should be prepared in detail by the construction manager or architect/engineer and reviewed with the owner very early in the design sequence. They should then be presented to the selected contractor as basic nonnegotiable requirements in the negotiations.

In competitive bidding, one of the major things that happens to the benefit of the owner is that the project is priced throughout the industry on a wide basis because a number of general contract bidders are each obtaining a number of prices from various subcontractors and material suppliers. Since the general contract markup for profit will normally run from only 3 to 6 percent and since it is not unusual for a given subcontract price or material price to vary by as much as 20 percent, it is extremely important to have the same degree of widespread competition from subcontractors and the various suppliers in a direct negotiation situation as one would have in a bidding situation. This is not altogether a comfortable situation for the selected contractor in a negotiated contract, because it is always an easier course, with certain side benefits to the selected contractor, for him to deal with favorite and selected subcontractors. However, it must be the responsibility of the construction manager or the architect/engineer to see to it that this widespread pricing does in fact occur. One procedure that can be followed is to first establish an upset price as an agreed maximum with the contractor. Then, an agreement should be reached on the contractor's fee, i.e., his profit for the project. Then, the selected contractor and the construction manager or architect/engineer can jointly receive competitive bids on major subcontracts and jointly receive prices on other materials and minor contracts. Subsequently, these can all be summarized and a lump-sum price or upset price, depending on the type of contract to be negotiated, can be reached.

Most of the foregoing discussion related to direct negotiation pertains primarily to those situations wherein a group II owner has chosen to negotiate with a contractor with whom he has never dealt before. There are, of course, those continuing relationships between contractors and owners that can be very advantageous to both parties. It is true, of course, that this situation usually exists because of the responsiveness of the selected contractor to the owner, and it is also true, in turn, that a number of provisions that have been discussed before, as well as a number of procedures that have been discussed heretofore relating to negotiation, can and probably should be modified in a given situation. However, when this situation exists, the contractor who does a major amount of continuing business with a given owner has probably actually assumed a number of the responsibilities of a construction manager and it is not likely that there would be a need for separate construction management services per se (i.e., separate from the services of an independent architect/engineer).

In the case of a negotiated contract, once the contract has been awarded—unless there is a special relationship between the owner and the contractor—there should be no difference between a negotiated contract and a bid contract in the approach of construction management during the construction phase. Chapter 12 deals with the construction phase—i.e., construction management services—and pertains equally to negotiated contracts and bid contracts for the several different basic management plans that have been outlined heretofore.

CHAPTER 12

Construction Management in the Construction Phase

It should be reemphasized that the term "construction management" means different things to different people; it is entirely correct to classify any or all of the activities covered in Chapters 5 and 7 through 11 as construction management. Yet, it would be just as proper to classify many of those activities as part of the design process, again pointing up the very necessary ties between design and construction management.

This chapter, however, deals with the most basic construction management activity, the common denominator in most construction management efforts. That is, the direct management of the construction operation or construction contract administration.

In order to deal with this basic phase of construction management, varying project management plans usually utilized in professional construction management services are divided into three categories, as follows:

1. Management and administration of a single general contract
2. Management and administration of multiple prime contracts
3. Direct construction management of separate trade contracts and work forces

In each case the discussion will deal with the phase commencing with the award of the first construction contract for the given project. Construction management services for previous phases are covered in preceding chapters.

(It is assumed that the owner is of group I.)

1. Management and Administration of a Single General Contract

The fundamentals of this fairly typical approach are that a general contractor has full and sole responsibility for fulfilling all requirements of the contract documents (drawings and specifications) for a predetermined price; the architect/engineer or construction manager makes a reasonable determination of that performance and recommends progress and final payments to the general contractor by the owner; in so doing, the construction manager or architect/engineer supplies certain responses and information to the contractor and reports to the owner.

Once the single-responsibility general contract has been awarded, the architect/engineer's or the construction manager's only real control is the purse string. Yet in our economy, this can be a most effective control. And the single general contract has the potential for the lowest cost and most effective administration for the owner.

It is true, of course, that management tools which have been earlier set in motion, such as critical path method scheduling, must be properly maintained and must continue to be fully utilized. But even these kinds of things, which are in the form of construction contract provisions, are effectively enforced in the final analysis only by purse string control.

The basic construction management services to be provided, once the construction phase has commenced, are not unlike the construction-phase services of the normal architectural contract as illustrated in AIA contract form B-131, the standard form of agreement between the owner and the architect.

However, some practices that have developed under similar contracts need to be reexamined, and some additional activities will be productive.

First, a reexamination of certain common practices:

The most unproductive and unwise practice that has developed in construction contract administration has been the use of the owner's representative at the site with a resultant division of responsibility and improper communications.

Under the typical contract between the owner and the architect/engineer, there are usually provisions for full-time representatives or inspectors to be assigned to the construction site. The unamended contract wording usually calls for these people to be provided by the architect/engineer on a reimbursable basis. The foregoing has frequently been modified so that these personnel are employed directly by the owner or are taken from the owner's own organization in the case of some municipalities and larger corporations. In other cases, inadequate numbers of experienced personnel have been authorized by the owner.

Taking some owners' view, this situation has developed because:

1. Personnel assigned by the architect/engineer were only journeymen, were not adequately supervised, were not familiar with the contract documents in advance (drawings and specifications), and were generally ineffective.

2. Owner felt cost would be lower by reducing force or supplying them from its staff.

3. Owner wanted "independent" observer at the site to watch the architect/engineer as well as the contractor.

The owners' view, expressed in item 1 above, unfortunately has all too often been fully justified. At the same time, this has been the result more of the typical contract between the owner and the architect/engineer and of the absence of an overall coordinated project management approach than of a lack of interest or expenditure on the part of the architect/engineer. In fact, the construction phase, under a typical owner-architect/engineer contract, has represented a break-even situation at best for the architect/engineer and has frequently seen substantial losses.

In the case of item 2 above, this has usually been a shortsighted move by someone "down the ladder" in the owner's organization, as the costs involved are almost negligible when compared to the total project cost, and this action has divided responsibility, often caused confusion, hurt design execution, and rarely improved construction quality, cost, or time control.

The case of item 3 above has represented, generally, a rather naïve view, on the one hand and created a most deleterious influence on the other. Good construction management is most likely to result from a single, coordinated team, providing both office and field services, justifying and having the owner's full confidence and support, and representing a single point of contract administration and project management throughout all phases of design and construction. Less than this is a compromise that is not likely to be in the best interest of the owner.

The architect/engineer or construction manager who has the responsibility for time, cost, and quality control must have an agreement with the owner that will allow him to do the following:

1. Have adequate personnel assigned during the construction phase at the jobsite, fabrication locations, test sites, and in the office
2. Utilize construction management personnel who are familiar with the contract documents, preferably having carried out a preissuance detailed review
3. Utilize construction management personnel who are regular staff members of the architect/engineer or construction manager firm or who have been brought onto the staff sufficiently in advance to have been trained in the specific type of contract administration and to have been oriented to the details of the project

Following are eight other areas of frequent failures, poor practices, or lack of management in the traditional administration of the single-responsibility general contract.

Failure to promptly condemn or reject work. Many times the construction manager or architect/engineer, in an effort to try to help a contractor out of a problem or due to inattention to detail, has failed to promptly and clearly condemn or reject work that did not comply with the contract documents. Thereby he not only sometimes may have hurt the owner's position and put himself in a bad position but actually caused the contractor more difficulties by not immediately letting the contractor understand clearly that the respective work does not meet the requirements of the contract documents. One point that should be borne in mind is that if removing a condemnation should later be deemed in the best interest of the owner, if the work is corrected, or if an alternative is found that is acceptable and in the owner's best interest, no undue damage has been done to either party in the interim. Some construction managers may hesitate to use such a strong word as "condemn" in notifying a contractor of what might be considered typically unacceptable work or minor deviations from the contract documents. However, it is best to utilize just such a concise term, so that the contractor cannot possibly be misled, and to use the same terminology throughout the project.

Failure to do plain, old-fashioned inspecting. The construction manager and architect/engineer would do well to bear in mind that inspection of the work is basic to construction management and that there is no substitute for it in this construction management configuration or in either of the two other categories discussed later in this chapter. It is certainly true that it must be made clear to the contractor and the owner that it is fully and solely the contractor's responsibility to comply with the contract documents whether or not inspection or observation of the work is carried out by others. Furthermore it is clearly not feasible to inspect every detail of fabrication, erection, installation, testing, and contract compliance. Nevertheless, good, thorough, ongoing inspection is fundamental to proper construction management. There is no substitute for an experienced and diligent inspector who has a good understanding of the contract documents, who knows what can and should be expected in the work, and who really carefully watches layout, preparation, installation procedures, workmanship, manufacturer compliance, and all phases of the construction process. An inspector or construction manager serves his highest purpose when he is able to prevent a departure from the contract documents or an error. This kind of experienced looking ahead is in the interest of everyone—including the owner, the contractor, the construction manager, and the architect/engineer.

Failures in contingency budgeting. In describing the characteristics and advantages of the lump-sum single general contract to the layman owner, many construction managers and architects/engineers tend to allow some misapprehensions to exist by not clearly letting the owner understand that all construction projects have a degree of unknown costs until final completion has been accomplished. Neither owners nor construction managers can expect either perfect contract documents or perfect site data. No such thing has ever existed. The architect/engineer surely has the responsibility to exert due care with professional competence, but the very best preconstruction drawing and specification preparation effort will still leave at least some minor errors and omissions. Some in-house owner's facilities managers, who should know better, sometimes take the amaturish position that the design and contract documents production should be perfect. They thereby delude themselves, unwisely mislead their management, and unnecessarily create an adversary relationship with their architect/engineer that is rarely in the owner's best interest.

The time to lay the groundwork to avoid this kind of misunderstanding is at the predesign or, at the latest, at the schematic design phase budgeting by setting up, with proper explanations, in-progress construction contingency budgets. See recommended project budget format in Chapter 7.

However, at the beginning of the construction phase, preferably just before the contract award, the construction manager or the architect/engineer should specifically draw the owner's attention to this budget item, again being certain that the owner understands its purpose and insisting that the funds be set aside. Note the suggested wording in the project budget recommended format for the contingency fund:

> In progress contingency fund to cover change orders for necessary adjustments to site conditions, minor design refinements, and correction of minor errors and omissions in the construction documents.
>
> (Note: The above fund is not for changes or additions to the Project. If the Owner desires a fund to cover such contingencies, said fund should be set up in addition to the foregoing.)

To summarize on this point, change orders on any project are altogether unavoidable. This fact, and the method of planning for change orders, must be clearly communicated to the owner in advance of the commencement of construction and the handling of them must be fully reported on throughout the construction phase.

Communications failures leading to loss of confidence by the owner. Nothing is worse for the owner during construction than a loss of confidence in the construction manager or the architect/engineer. Yet, this occurrence is so frequent as to be almost typical. The fact is that this occurs far more frequently as a result of poor communications than from a lack of effort or capability on the part of the construction manager or architect/engineer. For example, it seems beyond the comprehension of most lay owners to recognize the amount of effort put into a project on their behalf by the average architect, and therein lies a good bit of the problem. Better and continuous communications with all affected levels of the owner's organization is the only answer; and the fact is that this is very much in the best interest of the owner, not just a matter of better client relations for the professional. For when the owner loses sufficient contact with or confidence in the construction manager or architect/engineer, he is very apt to do things not in his own best interests, including complicating lines of communications with contractors. It is most important to keep a uniform and single line of communications from the party representing the owner to contractors during construction. Any other arrangement will not be in the owner's best interest and will often result in unnecessary opportunities for contractors to obtain contract time extensions and increases in the contract price. And the party representing the owner should be the construction manager or architect/engineer unless the owner is handling the entire construction management task internally.

Failure to deal within an overall strategy. The management of construction is like any other business management operation to the extent that there should be an overall and definable strategy. Once the strategy is set and understood, every move should relate to it until such time as a change in strategy is indicated. Whether the owner is of group I or group II has a great deal to do with the choice of strategy. The management plan, is, of course, fundamental to the strategy for a project. Other factors will vary from project to project and from owner to owner.

Misuse of field personnel. There are a number of common misuses of field personnel. And related to this, of course, is putting the wrong personnel on the project in the first place—only a warm body, inadequately trained personnel, or personnel without sufficient authority. One common misuse is to allow field personnel placed there to observe the contractor's compliance with contract documents to be "taken over" by the contractor. Many field people will allow this to happen without realizing it, while others even have the mistaken opinion that carrying out some of the contractors' supervision or attempting to solve the contractors' problems is a proper role for the construction manager's or architect/engineer's field personnel. What actually happens when the people in the field allow themselves to be used in this manner is that overall administration, observation, and objectivity tend to suffer to the detriment of the owner.

Another common misuse of field personnel has to do with insufficient authority for these people. Ideally, if design and principal communication will allow, total project administration, instructions, decisions, payment approvals, shop drawings, condemnations, and all other construction management activities during the construction phase should take place in the field. It is true that it takes a very large project to make this economically feasible, and even then it is not always possible—for a number of very legitimate reasons including the construction manager's or architect/engineer's organizational structure. However, in each case it is usually best to come as close to that ideal situation as possible.

Other misuses have tended to arise from the historical fact of inattention in selecting the personnel, inadequate funding for this phase of construction management, and inadequate attention to the planning of field activities.

Allowing the contractor adversary relationship to become a deteriorating influence on the project. The basic owner-contractor adversary relationship is usually an inherent part of most construction contracts, with the construction manager or architect/engineer usually acting on the owner's behalf. This is far from altogether unhealthy, nor can it really be avoided in other forms of construction management plans for group I owners. At the same time, a mature, unemotional, fair, uniform, and businesslike approach should avoid letting this adversary relationship deteriorate into unnecessarily bad relations of the sort that have often involved personal vendettas, unnecessary gamesmanship in tricky letter writing, wasted time, and a general sniping at one another. While the architect/engineer's or construction manager's first responsibility is to the owner, he should thereafter be as helpful as possible in every way to the contractor.

Failure to really utilize the contract documents. Once a project is well underway, it is true that more use is made by the general contractor, and

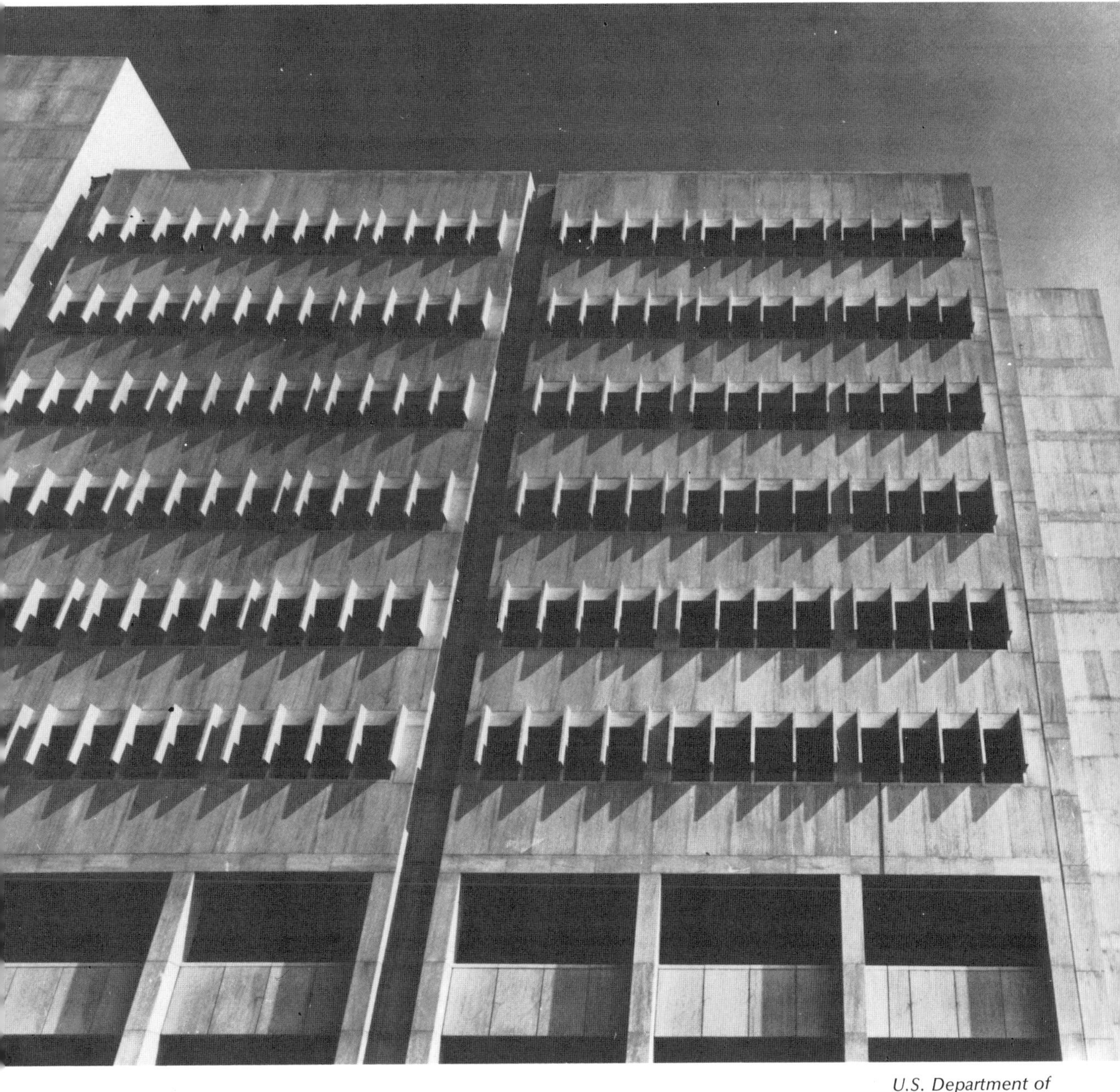

U.S. Department of Agriculture Research Laboratory, Athens, Georgia. (Heery & Heery, Architects & Engineers.)

particularly by his subcontractors, of the detailed shop drawings prepared for each phase of the work from the contract document drawings. In turn, it sometimes develops that almost no reference seems to be made to the contract documents by the contractor's forces at the jobsite. This situation must not lull the architect/engineer's or construction manager's field personnel into departing, themselves, from less than continuous and objective reference to the contract documents. If properly prepared from the outset in conjunction with the management plan, the contract documents should incorporate not only the needed technical information for construction but also all the enabling contract provisions for the respective project management plan. In turn, construction manager's and architect/engineer's personnel should make full use of them for both areas during all phases of the construction itself.

In addition to the foregoing, productive activities—particularly related to the control of time and cost—that can and should be brought to bear during the construction phase have been covered in Chapters 7, 8, 9, and 11.

Recommended Basic Procedures

Preconstruction Conference At the time of the award of the construction contract, whether it be the general contract or a separate contract, there should be a preconstruction conference including certain parties related to that individual contract. The conference should include the project manager of the architect/engineer or construction manager, a representative of the owner, a representative of the architect/engineer, and one or more persons from the contractor's organization including his jobsite superintendent and his project manager if the latter is other than the jobsite superintendent. If the project manager, first mentioned before, is not stationed full-time at the jobsite, then the construction manager or the full-time representative at the site of the architect/engineer should also be in attendance at the meeting.

The primary purpose of the preconstruction conference is to reconfirm the proper channels of communications and assist the contractor in understanding various methods and procedures that will help him in expediting shop drawings, requests for payment, and other paper work.

It is very important in this meeting to reconfirm the singularity of representation of the owner in administration of the contract as well as at the jobsite. If the architect/engineer is providing the construction management services, then it would be the architect/engineer who constitutes the point of contact with the contractor. If there is a separate construction manager, then it would be the construction manager who provides the point of contact with the contractor. In any case, the representative of the owner, who is present at the meeting, should be prepared to clearly reconfirm this singularity of owner representation. Any other arrangement or any confusion left after this meeting will not be in the best interest of the owner.

For projects that have contract completion dates or are to be expedited (which means almost all projects), the importance of time should be thoroughly discussed—along with the project schedule—in the preconstruction conference. On this occasion, the construction manager or architect/engineer should carefully draw the attention of the contractor to the various time-control contract provisions, such as payment scheduling, liquidated damage provisions, superintendent requirements, contract time-extension rulings, etc., that are contained in the general and special conditions of the contract. It will be crucial, at this meeting, for the construction manager or architect/engineer not only to point out these provisions but to be convincing to the contractor that he will fully administer the contract both fairly and completely and that he will keep a complete record of the job to allow him to do so in a proper manner. This kind of clear understanding at the outset can save the contractor as well as the owner a great deal of trouble later on in the project itself.

Contractor Payment Expediting One of the most important things that a construction manager or the architect/engineer for a project can do is to expedite payments to the contractor once the payment is due. Possibly of even more importance is for an owner, through his construction manager or architect/engineer, as well as for construction managers and architect/engineers on their own, to develop a reputation for quick payment once funds are due. This kind of reputation will not only improve the amount of bids in the first place but will generally improve the quality of services and work received from contractors.

Another factor in expedited payments for the contractor is that this effort helps offset any disadvantages of the owner-contractor adversary relationship, which is a necessary by-product of an arms-length construction contract.

There are others for whom the expediting of payments will bring about better prices and better services. They include testing laboratories, consultants, CPM consultants, and architect/engineers.

Prior to having the contract documents for projects issued for bidding, the architect/engineer or construction manager should learn of the payment procedures, through the owner's organization, that will be necessary for processing payments to the contractors and others for the project. To the extent necessary, he should go to the highest levels in the owner's organization to clear up any misunderstanding and to eliminate unnecessary procedures. Any particular procedures that will be required should then be described in the contract documents. Among the prior arrangements that the construction manager should make would be the arrangement for his organization to "walk" payment vouchers through the payment processing procedure. In order to obtain normal prices in the construction industry, a reputation for payment within ten to fifteen days after payment request approval by the construction manager or architect/engineer should be attained.

Decision Maker at the Jobsite It sometimes happens that, during the construction phase of an architect/engineer's services, the full-time representative at the site serves as no more than a warm body or at best a set of eyes and ears. This is unfortunate. If the resident representative is not given the authority that is necessary to properly and efficiently perform construction-phase services at the jobsite, he must go back to higher levels in the architect/engineers' organization to obtain decisions on almost every point. Personnel of this type, provided by either the architect/engineer or the owner, are almost worthless. As has been mentioned earlier, the chief representative of the owner at the jobsite should be from the architect/engineer if the architect/engineer is providing the construction management services, or he should be the construction manager. In any case, it is very desirable for this person to have the background, capabilities, and authorization to handle almost any problem that comes up in the field.

If the construction manager is separate from the architect/engineer, he will not generally want to approve any change orders, check shop drawings, or make any decisions affecting design. Except for this, the ideal arrangement would be to have the full construction management services take place at the jobsite office once the project has commenced. This is not always feasible on medium-to smaller-size projects; however, that situation should be attained as closely as possible in each case.

Change-Order Documentation The construction manager and architect/engineer should carefully avoid field orders, minor changes, no-cost changes, and trade-offs with the contractor without putting such in the form of change order.

The fact that a change order has no cost implication is no reason not to have it executed and numbered in the proper manner along with all other change orders. Extending footings, carrying piles to further depths, moving a door location at no cost, giving a contract time-extension ruling, changing an insurance requirement, providing the contractor with color schedule changes—all these should be documented in the form of change orders.

If any other procedure is followed, such as handling smaller no-cost changes with only field orders, later problems which will rarely be in the best interest of the owner will almost invariably develop.

Recommended Construction Management Personnel Staffing The following table shows recommended staffing of full-time construction management personnel for single general contract administration. The numbers given below should probably be considered as maximums in most cases and should also provide sufficient personnel for administration of minor separate contracts, such as landscape planting. This would include personnel assigned full-time to construction management once a contract has been awarded; while it would cover primarily full-time field personnel, it should also be sufficient to cover full-time personnel visiting fabrication and test sites and other related activities.

FULL-TIME CONSTRUCTION MANAGEMENT (CM) PERSONNEL FOR SINGLE GENERAL CONTRACTS

	No. CM persons required	No. clerical or apprentices required	Cumulative total No. persons required
Up to the first $2 million CCAP	1	0	1
Through the next $3 million CCAP	1	0	2
Through the next $5 million CCAP	1	0	3
Through the next $5 million CCAP	1	1	5
Through the next $15 million CCAP	1	1	7
Through each $20 million CCAP thereafter	1	1	9 or more

Progress Photographs and Periodic Reports to Owner's Top Management Monthly progress photographs are fairly routine with many construction managers and architect/engineers during the construction period. It would be appropriate for the cost of this to be reimbursed by the owner, and the requirement for the photography can probably be included in the special conditions of the contract so that, with the approval of the photographer by the construction manager or architect/engineer, the photography is actually provided by the general contractor.

The photographic record can be most helpful, frequently in very practical ways, later on in the course of work for making investigations, decisions relative to work that has been covered up, time-extension rulings, etc.

Another good use of a monthly photograph is as a means of reporting to the top officials or management in the owner's organization. Even though there may be an established procedure for the construction manager or architect/engineer to report to an identified representative of the owner, it is a very good idea for the construction manager or the architect/engineer to supplement this with a monthly report to the top management or official in the owner's organization. A brief summary of the progress and prognosis of the project—along with summarized cost data, status of contingency funds, and the job photograph—can be most helpful and should not be objected to as bypassing the regular representative as long as the regular representative receives copies of these reports and continues to be the primary point of contact with the owner's organization by the construction manager or architect/engineer.

Minor Separate Contracts and Delayed Purchases Even when the basic construction management plan calls for a single general contract, there may still be a few minor separate contracts or separate purchases. This may come about for a variety of reasons ranging from better cost control, expediting design phase, or better quality control.

Minor separate contracts might include contracts for landscape planting, signs and graphics, site clearing, and rough grading. Separate purchases or installation contracts might include carpeting, window blinds, appliances, and the like.

Cost control is one definite reason to consider separate purchases of items, particularly of a finished nature, that might have to be made late in the construction phase in a very large project whose construction span may cover 1½ or 2 years or longer. For example, in a $30 million high rise housing project with complicated subsurface and site conditions and a projected two year construction time, it would be much better to have the appliances purchased separately. The reasons for this are as follows:

1. Current model numbers at the time of completion can probably not be specified with confidence by the architect/engineer 1½ or 2 years in advance.

2. Since the purchase will be made by the contractor so far in advance, or since a nonconstruction item is involved, it will be necessary for the contractor to make a very conservative escalation allowance for purchasing anything so far beyond the date of contract award. (This would apply for other items such as carpeting, blinds, landscape planting, and appliances.)

3. Where there is little interface with the general contractor and his subcontractors on a very large project, frequently a better price can be obtained by the construction manager for the owner if a specialty installer/subcontractor or supplier can be dealt with on a competitive basis near the time of delivery requirement.

2. Management and Administration of Multiple Prime Contracts

There are a number of reasons, as has been discussed in Chapter 10 ("Phased, Separate, and Transferable Contracts"), for separate prime contracts. Several of these reasons, to review, are as follows:

- Several different buildings in a single project making it desirable, from a scheduling or other standpoints, to bid and award a separate general contract for each building
- The need to deal with major constraint activities by having early bidding and award of contracts for such items as structural frames, heavy mechanical and electrical equipment, elevators, escalators, bridge cranes, heavy site work, and foundations

- The use of preengineered industrialized building systems
- Projects under the jurisdiction of such governmental subdivisions as the State of New York, which requires separate contracts for general construction, mechanical systems, plumbing systems, electrical systems, and elevators as a minimum

On most projects in which separate prime contracts are used, whether or not these separate contracts are subsequently transferred into a general contract, the number of such contracts will usually range from around three to seven.

Basically, administering these separate contracts is the same task as administering the single general contract, as has been described above. Each of the separate contracts should be handled in the same way as the general contract, following the same recommended procedures with individual preconstruction conferences, contract documentation, etc., as was recommended above for the single general contract.

The additional responsibility for the construction manager or architect/engineer when separate contracts are utilized, is, of course, that of coordination between the contracts. This very substantial activity, including making rulings in the case of conflicts between two contractors, along with the added administrative and record keeping tasks for the multiple contracts for a given project size, will require additional construction management personnel.

The following shows recommended construction management personnel staffing for separate prime contracts:

FULL-TIME CONSTRUCTION MANAGEMENT (CM) PERSONNEL FOR MULTIPLE PRIME CONTRACTS

	No. CM persons required	No. clerical or apprentices required	Cumulative total No. persons required
Up to the first $2 million CCAP	1	1	2
Through the next $3 million CCAP	1	0	3
Through the next $5 million CCAP	1	0	4
Through the next $5 million CCAP	1	1	6
Through the next $15 million CCAP	2	1	9
Through each $20 million CCAP thereafter	2	1	12 or more

In conclusion, there are two very important points to keep in mind in administering separate prime contracts. One is that the planning for and the management of separate prime contracts must take place during and before preparation of the contract documents. As has been discussed in Chapter 10, the phasing, transferability of contracts, etc., must all have been worked out very carefully, starting with the management plan, and all the coordination and transfer provisions must be clearly covered in the contract documents for each and every separate contract.

The other point is a reminder on transferable contracts: The construction manager or architect/engineer should never require "acceptance" of a transferred contract. The owner's right to transfer one contract to another must be clearly spelled out in the documents of both contracts prior to their being issued for bid. Subsequently, when one is transferred to the other, there should simply be a registered notification of transfer without any request for "acceptance," as the latter would then imply that the contractor in question might have some rights, post bid, with regard to whether or not the contract can be transferred. This, of course, cannot be a question at this late date.

3. Direct Management of Separate Trade Contracts and Work Forces

This management plan, which can be most useful under certain unusual circumstances, is frequently erroneously referred to as "CM" or, even worse, "CMing" a project. Some, who are less than well versed in construction management, seem to have the misunderstanding that "construction management" is only employed or needed when there is no general contractor. Nothing could be further from the fact.

However, several of the unusual circumstances under which the direct construction management of separate trade contracts and work forces might be employed are as follows:

1. Complex renovation and alteration projects
2. Developmental projects
3. Very small, extremely high quality projects, where early cost knowledge is not a prime consideration
4. Special situations in which there is a lack of competition due to artificial circumstances
5. Very highly accelerated smaller projects with a series of early constraint activities

When this mangement plan is employed, it should be recognized and clearly explained to the owner that there is no general contractor in the conventional sense and that the owner is therefore giving up some fairly substantial advantages in the form of risks that a general or prime contractor normally assumes. Nonetheless, under the appropriate circumstances, the advantages of this management plan can be substantial enough so that, in evaluating the trade-off, the construction manager and the owner will conclude that this is the best approach in the individual situation. Under this approach, the construction manager takes over many of the roles normally played by general and prime contractors. These include direct supervision of forces and smaller subcontractors as well as the main subcontractors in the field. They include layout, direction of cleaning and miscellaneous labor forces, setting up field shanties (sometimes), handling insurance and safety programs, and usually all of those activities that are referred to as the "general conditions" work. The construction manager, however, does not normally undertake the legal risk and liability or surety positions of a general contractor unless there is that specific arrangement between the owner and the construction manager. In the case of the latter, the construction manager has then, really, become a cost-plus general contractor and is no longer serving the owner in the role of a professional.

Under this management plan, one would generally find somewhere between thirty and fifty separate trade contracts or field crew activities to be supervised and coordinated by the construction manager.

A great deal has been said regarding this construction management approach, which does away with the "adversary relationship" between the owner and a general or prime contractor. There can be some advantages along these lines under certain circumstances. However, if that is one of the aims of this approach, the construction manager and the owner should evaluate whether one or several "adversaries" are not being replaced by thirty or forty new adversaries. The latter could certainly be the case in many circumstances, to the detriment of the owner. A great deal has to do with personality and capabilities of the construction manager.

The accompanying table shows the recommended maximum staffing for full-time construction management personnel for the construction phase, using the management plan of direct construction management of separate trade contracts and work forces.

CONSTRUCTION MANAGEMENT (CM) PERSONNEL
FOR DIRECT CONSTRUCTION MANAGEMENT
OF SEPARATE TRADE CONTRACTS AND WORK FORCES

	No. CM persons required	No. clerical or apprentices required	Cumulative total No. persons required
Up to the first $2 million CCAP	2	1	3
Through the next $3 million CCAP	1	1	5
Through the next $5 million CCAP	2	1	8
Through the next $5 million CCAP	1	2	11
Through the next $15 million CCAP	3	1	15
Through each $20 million CCAP thereafter	3	1	18 or more

Index

Index